Lean Manufacturing:
A Plant Floor Guide

Lean Manufacturing:
A Plant Floor Guide

Edited by:
John Allen
Charles Robinson
David Stewart

Society of Manufacturing Engineers

Dearborn, Michigan

Copyright © 2001 Total Systems Development, Inc.

987654321

All rights reserved, including those of translation. This book, or parts thereof, may not be reproduced by any means, including photocopying, recording or microfilming, or by any information storage and retrieval system, without permission in writing of the copyright owners.

No liability is assumed by the publisher with respect to use of information contained herein. While every precaution has been taken in the preparation of this book, the publisher assumes no responsibility for errors or omissions. Publication of any data in this book does not constitute a recommendation or endorsement of any patent, proprietary right, or product that may be involved.

Library of Congress Catalog Card Number: 2001-132866
International Standard Book Number: 0-87263-525-2

Additional copies may be obtained by contacting:
Society of Manufacturing Engineers
Customer Service
One SME Drive, P.O. Box 930
Dearborn, Michigan 48121
1-800-733-4763
www.sme.org

SME staff who participated in producing this book:
Michael McLelland, Staff Editor
Walter Kelly, Consulting Editor
Rosemary Csizmadia, Production Supervisor
Kathye Quirk, Graphic Designer/Cover Design
Frances Kania, Administrative Coordinator
Jon Newberg, Production Editor

Printed in the United States of America

Contributing Writers
Total Systems Development, Inc.

John Allen is a principal of Total Systems Development (TSD), Inc. He has developed an international reputation as an expert in training on the Toyota Production System and lean manufacturing, and has published numerous articles on the human relations aspect of lean manufacturing. Mr. Allen is the former director of training for Toyota Georgetown, and helped launch the Georgetown Plant for Toyota.

Rod Centers is a consultant and project manager. Mr. Centers formerly spent three years helping Ford Motor Company define and implement the Ford Production System and was manager of stamping and welding operations at Johnson Controls, Inc.

Jim DeBold, a former consultant with TSD, has an extensive background in implementing pull systems. As a consultant, Mr. DeBold has supported semiconductor suppliers in the transformation to lean, and has assisted Ford Motor Company to define and implement a material management process in numerous manufacturing facilities throughout North America.

Richard Dixon is a consultant and project manager who has worked in a wide variety of plant settings. Mr. Dixon has more than 30 years of experience in manufacturing, including plant management positions.

Perry Hall is a consultant and project manager who has been a leader with General Motors Powertrain in its Competitive Manufacturing Group. Mr. Hall worked with Ford Motor Company to implement the Ford Production System and has eight years of experience with Toyota.

Mark High, formerly a consultant with TSD, worked with Big Three suppliers in implementing lean manufacturing. Previously, Mr. High served as a team production member in body weld and assembly shops at the Toyota Georgetown plant and was a production control specialist at Toyota Motor Manufacturing North America (Erlanger, Kentucky).

Ronald Link is a consultant and manager of TSD's Southwest Office. Mr. Link has more than 31 years experience in manufacturing-related human resources. He has in-depth international experience with site implementations, including a greenfield site in Mexico and a brownfield facility in Eastern Europe.

Richard Manley is a consultant assisting Ford Motor Company with lean implementation of its suppliers based on the Ford Production System. Previously, Mr. Manley was stamping assistant manager with Toyota Motor Manufacturing.

Anthony McNaughton is director of TSD Asia Pacific based in Melbourne, Australia. Mr. McNaughton works with automobile component and non-automotive manufacturers in the region. His background includes 10 years with Toyota. He is a Toyota certified trainer in standardized work and Kaizen.

David Meier is a consultant specializing in transforming non-automotive companies to lean manufacturing. Mr. Meier has worked with a number of leading companies, including Kodak, Hoffman Engineering, and Crane National Vendors. Previously, Mr. Meier was a certified trainer of Toyota's problem-solving courses and job instruction methodology during the 10 years he worked there.

Robert Mussman is an expert in lean manufacturing training and problem-solving. Previously, Mr. Mussman assisted Ford Motor Company in developing and implementing the Ford Production System. He also helped develop training programs and train the initial workforce at Toyota Georgetown.

John O'Meara, formerly a consultant with TSD, developed training materials and implementation strategies for lean manufacturing conversions. Previously, Mr. O'Meara helped lead the implementation of the Ford Production System in North America and worked with Toyota as a launch coordinator.

Dennis Raymer is a consultant who has assisted a variety of plants to implement lean manufacturing. Previously, Mr. Raymer helped lead the implementation of the Ford Production System at several engine and supplier plants. He also has developed training materials and implementation strategies for several other companies converting to lean manufacturing.

Rachel Renick is a consultant who has helped Tier One suppliers to Ford understand and deploy lean manufacturing concepts and tools. Prior to her duties at TSD, Ms. Renick worked at Toyota Motor Manufacturing in quality engineering and supplier development.

Charles Robinson is a principal of TSD and has developed a national reputation as an expert in Total Productive Maintenance and in developing corporate strategies with lean manufacturing. Mr. Robinson has co-authored a book entitled *Total Productive Maintenance: The North American Experience* (Productivity Press, 1995) and has published numerous articles.

Baikuntha Sharma, formerly with TSD as a consultant, specialized in process improvement. Previously, Mr. Sharma worked as a process engineer at Toyota and with the design and construction of manufacturing facilities.

David Stewart is a consultant and director of research. His areas of expertise include change management, executive coaching, and organizational development. Dr. Stewart has worked in a variety of for-profit and nonprofit private and public organizations, and has provided consulting and training services to IBM, GlaxoWellcome, UNC-Chapel Hill, General Motors, Ford, Daimler-Chrysler, Delphi, and AlliedSignal.

Gregory Thomerson specializes in implementing different phases of lean manufacturing. Mr. Thomerson has nine years of experience with Toyota Manufacturing, where he supervised production and new model development/introduction activities in the paint production department. He played a key role in the layout and setup of two Toyota paint departments, as well as introducing two major model changes and three new models.

Steve West is experienced in quality systems, problem-solving techniques, and day-to-day line management. Mr. West has helped Ford Powertrain implement the Ford Production System. He also assisted Delphi to implement its Quality Network Manufacturing System. Previously, Mr. West worked as a team leader, group leader, manufacturing specialist, and assistant manager for Toyota.

Table of Contents

Preface xv
Acknowledgments xix
Introduction xxi
by John Allen and Charles Robinson
   Core Values in a Lean Environment xxii
   Examine the Big Picture xxiv
   Principles of Lean Manufacturing xxvi

SECTION I: LEAN BASICS

1 Flow 3
by Richard Dixon
   What is Flow? 3
   The Impact of Flow 5
   Goals and Standards for Flow 6
   Identify the Current Situation 9
   The Creation of Flow 14
   Dealing with Discrepancies 18
   Information Communication 20
   Expected Results 21

2 Quality Feedback 23
by Perry Hall, Richard Manley, and Rachel Renick
   Define the Customer 23
   Inspection Standards 24
   Built-in Quality (Jidoka) 26
   Periodic Checks 26
   Communication 30
   Organizational Structure 37
   Problem Identification and Solving 37
   Conclusion 42

3 Lean Measurements and Their Use 45
by David Meier

Impact Areas 46
Are New Measures Needed? 46
Defining Measurements 49
The Measurement Process 58
The Measurement Cycle 60
Conclusion 69

SECTION II: LEAN IMPLEMENTATION

4 Mapping the Value Stream 73
by Perry Hall, Mark High, Anthony McNaughton, and Baikuntha Sharma

Current State Mapping 73
Future State Mapping 83
Creating the Future State Map 87
Conclusion 103

5 Business Case Development 105
by David Meier

Getting Started 106
Intangible Benefits 107
Plan, Prepare, and Do 109
Inventory and Materials 114
Manufacturing Labor 116
Manufacturing Equipment Maintenance 118
Material Handling Equipment 120
Other Costs 121
Customer Satisfaction 122
Employee Involvement and Morale 123
Safety 125
Determining Potential Savings 126
Building a Business Case for a Pilot Area 131
Cost of Implementation 132
Business Case Examples 141
Conclusion 151

6 Change Management in Lean Implementation 157
by David Stewart
Why Change Initiatives Fail 158
Factors for a Successful Change 159
Stages in the Change Process 160
Roles and Responsibilities 165
Resistance to Change 166
Conclusion 170

7 The Five-phase Implementation Process 173
by John Allen
Preparation 174
The Five-phase Model 175
Successful Implementations 183
Implementing Lean Elements 184
Conclusion 194

8 Creating a Lean Human Resource System 195
by John Allen
Mutual Trust 196
The Ideal Lean Human Resource System 198
Establishing the Ground Rules for the Relationship 203
Maintaining the Established Relationship 207
Managing the People System 213
Conclusion 215

SECTION III: LEAN TOOLS

9 Policy Deployment 219
by David Stewart
Principles 220
Advantages 221
Process 222
Implementation 227
Deployment Considerations 230
Conclusion 231

10 Work Groups 233
by Ronald Link

Waste Elimination 234
Assumptions About People 235
Work Groups Provide the Best Approach 238
Work Groups and the Workplace 239
Anatomy of a Work Group 240
Training 241
Who Trains and Who Gets Trained 244
Work Group Meetings 246
Where to Begin? 247
Steering Committees 252
Pilot Areas 253
Implementing Work Groups 254
Greenfield versus Brownfield 256
Conclusion 257

11 Visual Factory 259
by Gregory Thomerson

Benefits of Visual Factory 259
Visual Factory and the Elimination of Waste 260
How to use a Visual Factory System 261
Visual Displays/Visual Controls 266
Error Proofing 273
Implementing a Visual Factory System 274
Conclusion 275

12 Error Proofing 277
by Richard Dixon

Defects Versus Errors 279
Where to Apply Error Proofing 279
Phases of Error Proofing 285
Steps to Effective Error Proofing 287
Conclusion 290

13 Standardized Work 293
by Gregory Thomerson

Benefits of Standardized Work 293
Purpose of Standardized Work 294
Elements of Standardized Work 295

Standardized Work Documents 298
Preconditions for Standardized Work 301
Roles and Responsibilities 303
Conclusion 305

14 Quick Changeover 307
by Rod Centers and Steve West
How Much Time Reduction is Possible? 307
How Quick Changeover Helps 308
Implementing Quick Changeovers 310
Advance Preparation Illustration 314
Addressing the Issues 315
Streamlining Internal and External Activities 320
Conclusion 322

15 Total Productive Maintenance 325
by Charles Robinson
Definition of Total Productive Maintenance 325
Traditional TPM Steps 326
Mold the TPM Effort into a Lean Implementation 352
Maintenance Procedures and Standardized Work 354
Conclusion 355

16 Problem-solving 357
by Robert Mussman
The Problem-solving Mentality 357
The 5-Why Cause Investigation 360
The Investigation Phase of Problem-solving 362
Conducting a Thorough Evaluation 367
The "Do" Phase 370
The "Check" Phase 371
The "Adjust" Phase 373
Conclusion 373

17 Pull Systems 375
by Jim DeBold and John O'Meara
Just-in-time Defined 375
Traditional Material Management 377
Pull System Management 378

Grocery Store Example 381
Tools of the Pull System 383
Basic Rules for Operating a Pull System 400
Using the Pull System for Continuous Improvement 400
Expected Benefits 402
Pull System Considerations 404
Company Culture 407
Training 408
Ensuring Success 410
Conclusion 414

SECTION IV: LEAN APPLICATIONS

18 Greenfield Site Implementation 417
by Richard Dixon and Charles Robinson

Identifying Needs, Defining Concepts 418
Conceptual Design 418
Program Approval 420
Detailed Design Phase 421
Construction 430
Training 432
Startup 436
Ramp-up 441
Normal Conditions 442
Conclusion 443

19 Brownfield Site Implementation 445
by Richard Dixon, Dennis Raymer, and David Stewart

Emphasis on Change Management 446
The Lean Conversion of Brownfield, Inc. 447
Conclusion 484

Index **489**

Preface

Immediately preceding the publication of this book there was a softening of the economy that followed an extended period of unprecedented growth. During times of prosperity it is not a simple matter to convince manufacturing leaders and managers that they need to reduce waste, improve productivity, and cut costs. As economic growth slows, however, manufacturers begin to seek ways to improve their bottom lines.

Lean manufacturing concepts are excellent tools for managers charged with eliminating waste and improving profits. Although it involves an investment to make the transition to lean, the expected results can be compelling. A leading book on lean manufacturing asserts that converting a mass production facility to lean manufacturing doubles labor productivity throughout the plant while cutting production throughput times by 90% (Womack and Jones 1996). In addition, inventories can be reduced by 90% and lean processes can cut defective parts and scrap in half. Another source describes typical benefits from lean to be a 30% productivity increase, 90% work-in-process reduction, 50% space utilization reduction, 85% quality improvement, and 90% lead time reduction (Arizona Manufacturing Extension Partnership 1998).

The above figures have been confirmed by the authors' experience helping companies make the transition from mass to lean. But don't just take our word for it. An annual survey of 3,000 manufacturers dramatically makes the case for lean. The survey reports: "managers who said their facilities had either achieved world-class status or had made significant progress toward that goal were more likely to achieve higher finished-product first-pass yields, better on-time delivery rates, lower scrap rates, and bigger improvements over time to productivity, manufacturing cycle times, and a host of other measures when compared to plants that ranked themselves on a lower world-class rung." Plants making significant progress toward world-class status were two, three, or four times more likely to have implemented predictive or preventive maintenance, just-in-time production, quick-changeover techniques, cellular manufacturing, pull systems, lot-size reductions, and other lean manufacturing innovations. The survey results crystallized the benefits of lean: "Plants that have fully achieved world-class status have a median productivity (measured as value of annual plant shipments per employee) of $203,000 compared with $160,000 for the entire survey population, and $136,000 for plants that have made no progress toward that goal. Nearly 32% of plants in the two higher-echelon world-class categories (fully achieved and significant progress) increased their productivity by more than 20% in the last five years, compared with just 18.7% of the respondents who said their facilities have made no progress or some progress" (Jusko 2000).

It should come as no surprise, then, that interest in the lean production system has grown dramatically during the last decade. It is timely that the authors who have been involved working in lean facilities (most notably, Toyota Motors Manufacturing of Georgetown, Kentucky) and engaged in helping other facilities convert to lean should share their practical and experience-based knowledge. All of the associates of Total Systems Development, Inc. have benefited from authoring this book. We have been able to capture some of what we know and thus increase our intellectual capital. Our company has demonstrated that we can conduct lean implementations and share our knowledge of how to do it. The growth of our company is an indicator that the popularity of lean manufacturing techniques and concepts is on the rise.

This book is intended for those who want to know about both the "what" and the "how" of lean implementation. It is not exactly a how-to manual; rather, it documents the implementation process more in terms of "what" than "how." Since each organization has its own unique circumstances and culture, the process will differ from one to the next. This book puts into place what is relatively common in most lean conversions. Every author has experience in lean implementation, and each chapter serves as a real and practical guide for those charged with implementing lean in their facility, whether they are team leaders, plant managers, floor supervisors, etc. A careful reader will gain a significant understanding of the pragmatic issues that must be considered in the complex undertaking known as a lean implementation.

This book is divided into four primary sections. The first section covers the basics of lean. In the introduction, we describe the five principles of a lean implementation that are critical to success. The concept of flow, which is the core of the lean production system, is described in Chapter 1. Understanding the nature of flow and how it is produced in the production system will enhance your ability to grasp the nature of the tools and the implementation process described in later chapters. Chapter 2 involves developing a quality feedback system—critical to maintaining flow. The last chapter of the section is on lean measurements and their usage. Knowing how to make use of metrics in a lean implementation is an important part of a successful implementation.

The second section details the process of the lean implementation. Chapter 4 describes the current and future states, and the subsequent mapping process that heralds the beginning point of the implementation. Chapter 5 informs the reader how to develop a business case to justify the initiative. Knowing how to make the case for lean becomes an important part of the change management strategy described in Chapter 6. Perhaps the most important chapter in the book is Chapter 7, which details the five-phase implementation process. This chapter gives structure to the process that is the most successful in lean implementations. Chapter 8, no less important, informs the reader of how to create the human system organization that supports the value-adding processes.

Lean tools are described in the third section of the book. Although there are many potential lean tools, we settled on nine we believe are the most important. Perspectives on implementing policy deployment, work groups, visual factory, error proofing, standardized work, quick changeover, Total Productive Maintenance, problem-solving, and pull systems are described in Chapters 9–17. Practical suggestions are included, as well as how they fit into the five-phase implementation process.

The last section, on applications, describes lean implementations in greenfield and brownfield settings. Although each implementation is unique, the reader will gain insight into approaching his or her own lean initiative by reading about some of the considerations in these two applications. Chapter 19, on brownfield implementations, is a must-read for anyone responsible for bringing lean into an existing facility. A case study in Chapter 19 is a practical example of a typical implementation, providing details that can be useful as the reader considers his or her own situation.

David Stewart
Hillsborough, NC
2001

REFERENCES

Arizona Manufacturing Extension Parnership. 1998. *Transforming the Way You Do Business*. Scottsdale, AZ.

Jusko, Jill. 2000. *Paths for Progress*. Industry Week, December 11.

Womack, James P. and Daniel T. Jones. 1996. *Lean Thinking*. New York: Simon and Schuster.

Acknowledgments

This book is the product of the combined efforts of many different individuals. The chapter authors were the largest contributors. Their efforts to share their practical knowledge will benefit countless readers. Others helped by editing, proofreading, and offering suggestions and constructive criticism. Specifically, Steve Avery, Rick Dixon, David Meier, John O'Meara, Charlie Robinson, and Baikuntha Sharma contributed time and effort to review the chapters. Thanks to Deborah Prechtl for her clear and interesting illustrations. Larry Bauman has helped keep us honest about our sources. Thanks to Michael McLelland at the Society of Manufacturing Engineers for his editing prowess and help to bring structure to a challenging undertaking.

Introduction

By John Allen and Charles Robinson

For the past seven years, the authors of this book have been involved in converting manufacturing facilities and their organizations from traditional mass production to an innovative production system known as lean manufacturing. *Lean manufacturing* is a term first used in the book *The Machine that Changed the World* by James Womack and Daniel Jones, which describes the manufacturing philosophy pioneered by Toyota. Lean manufacturing is practiced at Toyota under the name Toyota Production System. This book explains how to develop, evaluate, and implement lean manufacturing.

The authors have devoted much time and effort toward helping manufacturing facilities eliminate all forms of manufacturing waste from their sites. After much trial and error, we have developed a process that has a high probability of success. This book is about that process. It was written to assist manufacturing professionals planning their own improvement processes, and to share our experiences so others can learn from our successes and failures. We feel that implementing lean manufacturing is necessary for facilities to remain competitive in today's worldwide marketplace. Those who embrace the concepts of lean will be able to remain competitive and gain their share of the market.

Lean manufacturing, in a nutshell, is the endless pursuit of eliminating waste. Waste is anything that adds cost, but not value, to a product. In general, there are seven wastes in manufacturing, which include:

1. defects—producing parts that fail to meet product specifications;
2. waiting—people or operations waiting because of lack of material, equipment, or information;
3. motion—the movement of material, equipment, or personnel that does not add value to the product;
4. over-processing—performing operations not required to manufacture or assemble the product;
5. over-production—making more product than the customer demands;
6. inventory—excess raw material, work-in-process, or finished goods inventory; and
7. inefficiency—people wasting time, efforts, or ideas, equipment waste in capacity, or using more material than is required to complete the job.

Although these forms of waste seem rudimentary, many companies do not practice eliminating them on a regular basis because they often get confused in the details.

CORE VALUES IN A LEAN ENVIRONMENT

Some core values must not be violated in the pursuit of lean manufacturing. Project managers and operators must always consider core values as they develop and build a lean facility. Violating them seriously inhibits the probability of success. The core values are:

- job security;
- problems are good;
- floor-level involvement;
- value-add/support value-add; and
- accountability.

Job Security

Lean manufacturing does not mean eliminating jobs. The lean implementation effort must involve all employees in the organization, especially those who add value to the product. If jobs are eliminated due to implementation, progress will be slowed or stopped. If managers and top-level executives guarantee job security, progress will accelerate even though efficiencies are gained through lean manufacturing. That is why job security (Chapter 8) is a core value.

Problems are Good

The only way issues are going to be addressed is if they surface. The quickest way to stop progress on lean manufacturing is to hide problems. Therefore, those who bring problems to the surface should be praised and not punished. Problems are good—they are opportunities to make improvements (see Chapter 7).

Floor-level Involvement

Lean manufacturing is based upon the premise that those who are closest to the work have the best opportunity to develop innovative improvements. It is necessary to involve floor-level employees when improving a business' performance. To do this, the needs and expectations of the business must be communicated to the plant floor level using the goal-setting portion of a robust policy deployment process (discussed in detail in Chapter 9).

Value-add/Support Value-add

There are only two types of jobs in a manufacturing facility—those that add value to the product and those that support the people who add value to the product. This philosophy simplifies all aspects of roles and responsibilities. The most important employees in the organization are those who add value. Organizations that forget this core value are inefficiently using the most important resource in any company. (For a detailed discussion of human resources in a lean effort, refer to Chapter 8.)

Accountability

It is the authors' premise that, for the most part, people want to succeed, they want to be treated like adults, and they want to know the whole truth (whether the news is good or bad). If this premise is true, then it follows that those who add value want to be held accountable for the results of their efforts. Accountability means two things:

1. The goals of the organization must be deployed to all levels of the organization.
2. Progress toward goals must be regularly reviewed with all who are accountable for the results. Often, this area is either left out or minimized. The review should include positive reinforcement for on-track performance, as well as countermeasures for goals not on track.

The authors have learned these values from hard-won experience. In the past seven years, our company, Total Systems Development, Inc., has been involved with many manufacturing facilities in various sectors of the economy. Our involvement has ranged from advising to consulting to directing lean initiatives. It is not exaggerating to say that we have saved some companies from extinction. For all, we have helped reduce costs and improve efficiency. Our developmental efforts have helped those who lead to carry on after we have left. Our intention is to teach, and, by doing so, work ourselves out of a job.

Implementation is a tricky business. Often, the formula for success at one site is neither appropriate nor sufficient for another site. Many facets of lean implementation must be managed to achieve the lean sponsors' goals of creative implementation and sufficient transfer of knowledge. Some of the facets include the flow of material and processes, as well as support processes (for example, information systems, human resources, and finance).

EXAMINE THE BIG PICTURE

No issue in the enterprise is too small or unimportant. To achieve the best results available, the entire system must be managed. Overseeing and directing the whole system may sound imposing; indeed, it can be if it is taken as a series of separate initiatives. Implementation can falter when issues are raised but never solved. Change initiatives are usually treated as independent entities, and are not joined with concurrent system-wide problems. Ultimately, the solution is to ensure that whatever the level of implementation undertaken, lean leaders and implementers have the primary responsibility to integrate issues.

The suggestion that the person adding value to a product is the problem is absurd, and the notion that management can solve all issues can seem plausible. The issue is not who is right, but how the energy committed to lean ultimately can solve both of these issues. Managers must take the lead in securing support, and value-adders must take primary responsibility for transferring the change to their jobs. The bridge for both of these is awareness and training. Leadership can either pay lip service to the implementation, or it can prepare the organization and train employees for success.

Lean implementation is all-encompassing, spreading to finance measurement systems and information technology support, so even support organizations must be involved in the implementation. Companies that choose to ignore this will spend inordinate amounts of time trying to make up for their shortsightedness.

All organizational issues that arise need to be interpreted through the lens of lean manufacturing. The fundamental question that arises when new proposals are considered, will, by necessity, be "is this consistent with lean manufacturing principles?" Thinking in isolation is dangerous for both the organization and the permanent implementation of lean. Problems need to be elevated—thinking and planning must precede action. Personnel will learn to succeed by achieving support from others, and not by isolated action that results in individual recognition.

Lean manufacturing is not about the infusion of technology into a manufacturing process, especially if it is not directed toward assisting the operators adding value. We see many industries attempting to increase manufacturing efficiencies by adding sophisticated material resource planning systems (MRP) or enterprise resource planning systems (ERP). Few, if any, of these expensive system implementations are truly successful in reducing total costs. Most simply add an additional buffer between layers of the organization and effectively stifle communication, creativity, and sense of control for those adding value to the product. Unless the technology directly supports value-adders, it usually brings waste to the system. In fact, we suggest using the low-tech approach of pencil and paper for a number of evaluation techniques throughout this book to get a hands-on feel for that particular process.

PRINCIPLES OF LEAN MANUFACTURING

The lean implementations the authors have completed would not be successful if it were not for a number of principles developed from experience. These principles have been derived from years of hard-won battles while trying to implement lean manufacturing. There are only five, but their implications are the essence of the production system. At each step of implementation, managers should pause and reflect on how they are being upheld.

Having a simple set of principles to follow is useful and necessary to sort through all the issues that will arise. The following principles are not the only benchmarks for the process; however, they are the most important when undertaking a system-wide, systematic implementation of lean manufacturing.

The Customer Defines Value

Value is that for which the customer is willing to pay. It is important to realize that value is what the customer deems it to be. Often, the most important conversation is between the customer and the producer to determine what is and what is not considered value. Once this has been discovered, the producer can more accurately ensure that production processes are consistent with the customer's desires. Many times, value (from the customer's perspective) is assumed but turns out to be wrong.

When producers assume they know what the customer values, they risk creating a product that is fine from their perspective, but completely unsatisfactory in the eyes of the customer. When that happens, significant cost and risk have been put into essentially creating waste. For a business, it can spell disaster. When the customer's perspective is clarified at the beginning of the process, producers can examine their processes and determine what portion is value-added compared with non-value-added content. The goal is to increase value-added content and decrease non-value-added content.

It is critical that the production process includes the customer's values. Each operation must be examined to determine how it is intended to add value. Working backward, every job must be supported by the previous process so the final product is satisfactory to the customer.

Producers achieve customer satisfaction in each workstation. The strategy of maintaining product quality by final inspections and repairs is not only expensive, but makes the process of finding the root cause of problems much more difficult. In addition, inspection and repair are non-value-added and should be eliminated, or minimized, as much as possible. The only reasons for a detailed final inspection are if the manager does not believe quality can be maintained in the process, or if the customer believes it adds value and is willing to pay for it.

The Customer Establishes Pull for the Production Schedule

Production planning should be based on creating only what the customer wants. As a result, the rate of unit production becomes equal to the rate of consumption by the customer. If the goal for final production is the same as the customer's consumption rate, then the production rate for each preceding process also should equal that consumption. Lean implementers must establish that ideal rate, then make sure each process has value-added content equal to that rate. However, very few processes only have value-added content—most have non-value-added content either equal to or greater than the value-added.

The goal of each process should be to make only what is needed by the customer, when the customer wants it, and in the proper amount. This requires that a level of precision be patterned for each cycle, which is the best way to ensure that the customer is satisfied. This pattern is called standardized work (for a detailed discussion on standardized work, refer to Chapter 13). It is different from old work standards in that the operator determines his or her own work, then manages improvements to the process. When the customer changes the rate of demand, operators work together to rebalance the line so the new rate is met. Ownership taken by operators from this process is a major factor in sustaining pull by the customer.

Empower the People Adding Value

This is the most powerful, yet the most difficult, principle to achieve. The difficulty lies in dealing with people, who can be most unpredictable. The power of this principle is in unlocking the creativity and synergy of a group of employees who are focused on improving the business. A business focused on lean must make sure the evolution of involvement and empowerment is consistent with creating and maintaining a lean organization.

When establishing initial lean practices, the actions of those involved must be disciplined to achieve a foundation in precision. This requires individual action and accountability, not teamwork. Teamwork is necessary only when it is time to rebalance job content for customer pull. Up to that point, individual action is sufficient. As the five phases of lean progress, the necessity of having empowered, synergistic team members becomes greater. Empowerment and involvement is the destination, not a point of departure.

Preparation is critical because empowerment and involvement require that individuals be prepared to work in a group. Training and real-life experience on the job are necessary to reach the appropriate level of preparedness. Training in problem-solving, meeting facilitation, conflict resolution, and group dynamics is a foundation for effective group interaction. Experience on the job comes from solving problems, meeting to convey information, and resolving differences between team members.

If the importance of value-added work and eliminating non-value-added content is considered, the most important employees in lean manufacturing are those who make the products every day. Everyone else in the organization is designated to support that process. Value-adders become the focus of empowerment. They are the best prepared to offer ideas about improving value-added content through waste elimination.

Value-adders should make decisions resolving problems and issues, but this is possible only if they have the information necessary to make educated decisions. Value-adders need information about goals for the organization, objectives of the components needed to accomplish those goals, and accurate information about the current level of performance. This allows them to be productive employees who participate in the improvement process.

For most organizations to survive, they must continue to improve. To create an environment for improvement, it is necessary that people be treated with trust and respect, be treated like adults, and be allowed to do their best work. Managers who create this kind of organization must have a different mindset about employees who add value. Currently, many managers believe that a highly motivated staff is desirable, but the organization should focus on pushing product through the system. The present focus is on the volume of production, not on processes and contributions the operator can make. The basis for the lean mindset is that managers believe everyone can contribute; development of the human resource is basic to business performance; and everyone is in the process of eliminating their jobs.

The conditions whereby employees are motivated to eliminate their jobs are simple. Improving the company is the critical link within his or her control to improve job security. Employees participate in improvement activities primarily if their work will result in a better situation, such as participating in a full-time continuous improvement team, accepting a cross-functional assignment, or transferring to another desirable job. The absence of this link will make the employee feel his or her job security is threatened.

Eliminate Waste in the Value Stream

If the organization can deliver value to the customer, it stands to reason that eliminating non-value-added activities is fundamental. Ultimately, waste is eliminated due to successfully applying all the lean principles.

The goal of lean manufacturing, however, is not to install pull systems or reduce changeover times. These are important goals in the implementation process, but are secondary to the overriding goal of eliminating waste. Successful lean implementation occurs only when waste has been eliminated from the value stream. Other ways to reduce waste to zero must continually be sought. Tools are only a means to this end. Facilities may have Andon boards and use Kanban cards, but that is no guarantee that they are lean.

Use Total System Cost to Drive Performance

The mission of manufacturing is to minimize total system per unit cost while maximizing quality and safety and minimizing response time. The lowest total system cost is achievable only if safety and quality are maximized and response time is minimized. Traditional performance measurements focused on the pieces rather than the whole. The result is that managers can no longer see the forest for the trees. Effort is put into optimizing specific functions and components, without any sense of the impact on the cost of doing business. When the wrong things are being measured, we no longer know whether progress is being made.

It is not an easy matter to shift to a total system cost measure. New measurements, such as first-time yield and total manufacturing cycle time, have to be put into place and new systems for collecting data must be instituted. There must be a collective effort to emphasize how important the new measurements are, and reinforce and reward performance based on them. Advantages become apparent once problems have been worked out and new performance measurements are in place. Project managers then will know the relative contribution of different components to the product's cost. In addition, they will see new opportunities for eliminating waste and encouraging better performance.

Lean principles are simple, but they are difficult to implement. Difficulties result from trying to interpolate these principles into the existing system. Re-engineering an old system to look like lean, but not act like lean, is a common mistake. For these principles to result in a lean operation, implementers must start with fundamental decisions and re-examine them in the light of lean principles. Lean principles will lead you to a different place than where you are currently familiar.

To use this book effectively, you must assert the same discipline that will be required for implementation. First, read the book in its entirety. Don't expect to completely "get it" at first. For most of us who have worked in a company like Toyota, it took several years of daily implementation to be able to connect actions with the principles involved. It was several years after that before we could predict what principle would come into effect at any given time.

Once the concepts are familiar, focus on the five-phase implementation process (Chapter 7) and write a work plan for your facility or department for a six-month period. Once it is written, the work plan can serve as an expectations document for everyone to connect to the changes taking place. Additional work plans will need to be developed for after the initial six-month period. The key to success in lean implementation, however, involves the people, not the tools or technology, undertaking the implementation. Therefore, you should constantly check to see if the expectations set by the work plan are being met.

Good luck in your lean journey. Don't try to go it alone. There is plenty of help around. Use outside resources and insist they transfer their knowledge to you so you can carry on when they are not around.

Section I:
Lean Basics

Flow   1

By Richard Dixon

Efficient flow should be the natural state in a manufacturing environment. Why should material have to stop after each operation, get loaded into some type of container, be moved by a forklift to a storage location, and then moved by forklift again to the next operation? This start/stop cycle repeats until the product is complete, unnecessarily wasting time and resources. The *start/stop process* typifies a mass manufacturing environment—forklifts and racks with significant waste in the movement of material, floor space for stores, and more time needed to get the product shipped.

The *flow process*, by comparison, moves material via a short conveyor from one operation to the next. The only forklifts required are for loading raw materials at station 01 and for shipping the finished product. Figure 1-1 illustrates both flow and start/stop processes.

Which illustration best represents your business, Figure 1-1a or Figure 1-1b?

In a manufacturing environment, the objective is to ship the best quality products, on time, at a reasonable price (beating the competition). Costs go up when products are moved sideways, making the company less competitive.

WHAT IS FLOW?

Flow is the continuous movement of material through the manufacturing processes and on to the customer. If an operation with perfect flow could be established, nearly all waste associated with inventory, storage space, and transportation, as well as a great deal of waiting, could be eliminated. All other things being equal,

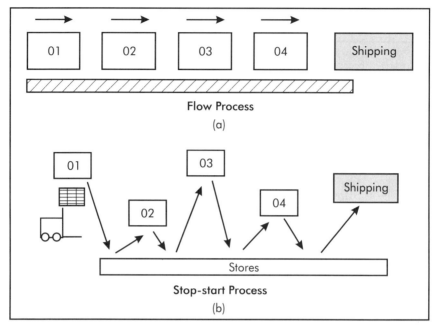

Figure 1-1. Flow and stop-start processes.

flow processes have a major positive impact in two areas—cash flow and inventory. (Other issues that are not equal in flow and mass environments will be addressed later.)

How is cash flow affected? Efficient flow processes move the product to the customer much faster, resulting in receivables being collected in a shorter period of time. If the company gets paid faster, that money can be invested and receive interest, or the company may not have to borrow funds to finance short-term operations.

Inventory is the only financial asset that is also a physical and financial liability. It is an asset on the company's balance sheet, but it is an operating liability. More negative issues regarding inventory can crop up than positive ones. In fact, other than being sold or converted, inventory will likely create several negative results:

- Inventory becomes lost.
- It becomes physically damaged by movement from one location to another.
- An engineering change makes it obsolete.

- It gets scrapped.
- It has to be reworked.
- Lost inventory results in extra setups.

THE IMPACT OF FLOW

Flow is in all processes—some companies have very little, others a great deal. In manufacturing environments, the amount of flow goes a long way to determining the profitability of a company.

Exceptional flow leads to:

- lower inventories;
- better quality;
- less floor space;
- better communications;
- quicker responses to problems, and
- faster throughput.

Lower inventories. Good process flow means inventory is moving continuously and not collecting between processes or in storage locations. The end result is a lower inventory investment.

Better quality. In a good flow process, each operator (a customer) is using the part that was just produced by the previous operation. If a part has a quality problem, the operator notifies the previous operator and production is stopped until the quality problem is corrected. The results are less rework, less scrap, and fewer defects to the end customer.

Less floor space. If in-process inventory does not exist, allocated space between processes is not required.

Better communications. Processes are positioned close together so operators are close to each other and can communicate easily.

Quicker responses to problems. Quality problems are discovered as they occur, not three days after the problem. The problem is confined to the process and not the resulting batch of inventory produced in a mass environment.

Faster throughput. Because the material does not move between processes and stores, it travels in a straighter line and gets to the shipping point much faster. Flow allows faster responses to customer changes because as the customer's order changes, the processes are changed to meet the new demand. Large batches of product are not sitting around partially completed through each process.

Companies still processing in large batch and queue mode always seem to be under the gun to get orders completed on time. Some companies have rack and forklift investments greater than the gross domestic product of a few small countries. How do they survive? They require larger facilities, a larger investment, more equipment, storage space, rework areas, repair operators, inspectors, and forklift drivers. They survive because they have become efficient at inefficient production. As more companies change to efficient flow processes and the competitive advantages they bring, the mass manufacturing dinosaurs will have a tough time surviving.

For the antithesis, consider the example of Toyota Corporation, which many consider to have the best flow processes in the world. In 1997, Toyota's flow processes gave it a minimum $400 per car competitive advantage over the Big Three automakers in the U.S. Are the Big Three focusing on developing flow processes in their own companies? Absolutely, and they are making great strides, albeit slower than they planned.

Toyota upped the ante with the 2000 model year Camry Solara™. Toyota's Solara plant in Canada is shipping cars five days after receipt of the order. This stunning advancement comes when its competitors have 50–90 days' worth of vehicles sitting on lots. The Big Three currently take 28–48 days to ship an ordered vehicle. On high-demand vehicles, the wait can be as long as six months. And the Solara ships in five days!

GOALS AND STANDARDS FOR FLOW

Companies that lack the competitive advantages of flow processes are looking at a dim long-term future. It is imperative that companies begin converting to flow processes now. Converting to flow is similar to undertaking any lengthy journey—it begins with a single step, usually by developing a plan to achieve the flow process goals.

During the planning stage, the following lean practices merit strong consideration:

- pull systems;
- one-piece flow;
- cellular operations;
- product orientation, and
- balanced operations.

Pull Systems. Pull systems help establish the flow operating philosophy. The basic premise says not to make a part until the next operation needs (pulls) it.

One-piece flow. The basis of flow processing is that one-piece flow is predominant in a facility's operations. This means that parts do not collect between operations. Each operation in the process is working on the next part for the following operation.

Cellular operations. Operations must be tied together in work cells. The best use of floor space is embodied by U cells, but S cells and straight cells sometimes give similar results. For example, a three-person assembly cell makes sense as a straight cell. (Cell shapes are discussed in greater depth in the section on creating flow.)

Figure 1-2 shows a cellular operation with balanced, one-piece-flow. Figure 1-3 is a mass manufacturing representation. In Figure 1-2, operators are working in a U-shaped cell with counter-clock-

Figure 1-2. U-shaped cell.

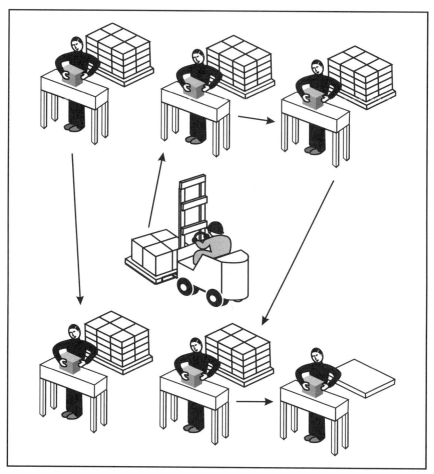

Figure 1-3. Mass operation.

wise one-piece flow. If problems develop, communication is quick and effective. In the mass environment of Figure 1-3, material builds up between processes, increasing inventory investment and the likelihood of quality problems.

Product orientation. Pull systems are impossible when equipment is grouped together in similar processes. A cell should include the processes required to create the product. Flow is about product orientation, not process orientation.

Balanced operations. One-piece flow requires that operations be balanced. With person A feeding parts to B, who feeds parts to

C, the process times for each operator must be close so that waiting (waste) is eliminated.

IDENTIFY THE CURRENT SITUATION

Before plans to reach a destination are developed, lean implementers must know their starting point. A current state assessment, also known as a current state map, will identify the producer's existing situation. Figure 1-4 shows an example of a current state map. (The mapping process is covered in detail in Chapter 4.)

The map is a graphical representation of a facility's process flows. The flow boxes represent each function in the process. Large arrows are used to connect the flow boxes, showing whether a pull system is in use. External partners (suppliers and customers) are represented by boxes as well, but have "rooster peaks" as a top line. Tombstones represent inventory, and data boxes identify metrics of the operations. Small arrows represent communication flows. Figure 1-5 shows an example of a future state map (also covered in Chapter 4) with much of the in-process inventory eliminated.

One key measure in a lean environment is dock-to-dock time. In the current state map example, it took roughly 257 hours to convert raw material to shipped product. In comparison, the future state map example reduces that time to just over 32 hours. This was accomplished by rearranging equipment to produce one-piece flow.

Figure 1-6 illustrates some obstacles to beware of as an operation is converted to one-piece flow. In this "sea of inventory," the manufacturing process is the ship. The inventory is represented by the water level. As inventory is reduced, the water level drops, exposing the rocks of the seven wastes. Each reduction in the inventory level exposes another rock. Each rock must be broken up prior to the next reduction in inventory. When all the rocks are gone, effective one-piece flow is achieved.

A critical element to identifying a facility's current situation is the ability to observe all the details impacting production flow. If an employee has worked in the same facility for a number of years,

Lean Manufacturing: A Plant Floor Guide

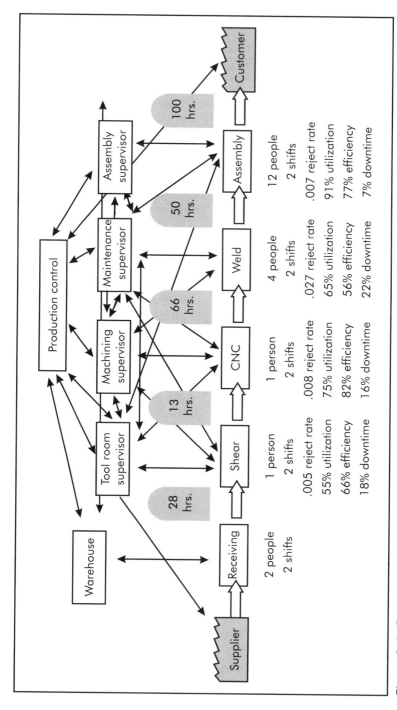

Figure 1-4. Current state map.

Flow

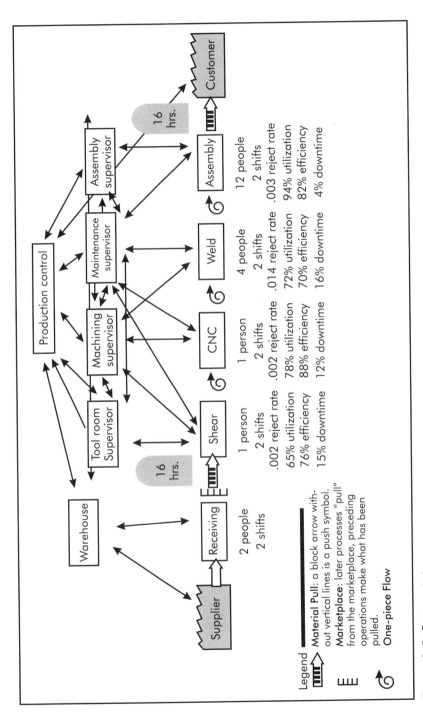

Figure 1-5. Future state map.

Lean Manufacturing: A Plant Floor Guide

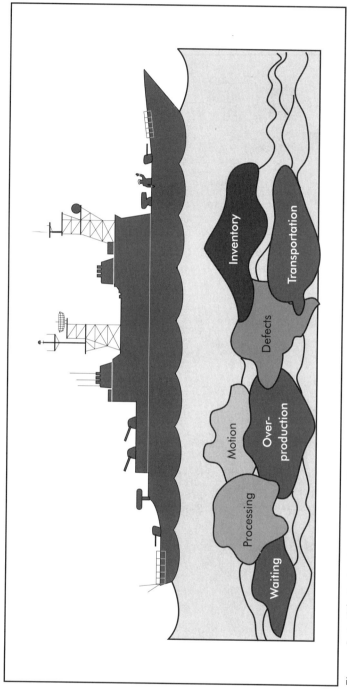

Figure 1-6. Obstacles to one-piece flow.

his or her observation powers are limited by that environment. To that employee, the current state of the facility represents the right way to make the product.

To accurately assess the current situation, all prior knowledge of the current facility operations must be eliminated. The slate must be clean—the person(s) responsible for implementing a lean program within their facility must develop new eyes that notice everything about the operations, and in great detail.

Corporate powers of observation leave a lot to be desired. For example, in the mid-1990s a customer was having delivery problems with a supplier. The customer scheduled a visit to the supplier's plant and asked for various data to be collected, including the number of steps involved in converting raw material to shippable product (units were not being shipped on time). During the meeting with the supplier, the discussion included operational steps in the manufacturing process. The supplier identified 17 steps that were required to convert the raw material to shippable product. "Impossible," replied the customer, "there must be more than that."

A trip to the floor with a video camera proved that there were actually more than 80 steps in the process. The supplier had only counted the value-added steps, not the 66 nonvalue-added steps in the process. This eye-opening event led the supplier to improve processes significantly by focusing on the nonvalue-added operations. It converted its processes to flow and got rid of the waste.

Indicators of waste include:

- racks,
- long conveyors,
- like machines grouped together,
- setups > nine minutes,
- inspectors,
- physical inventory,
- forklifts,
- baskets for in-process items,
- distance between operations,
- cardboard containers,
- lack of teams,
- operators tied to one machine, and
- everything that does not have a dedicated location.

THE CREATION OF FLOW

Once the successful evaluation of a company's existing facility is completed, and all the potential areas of waste have been identified, the next step to creating efficient manufacturing flow is design. Process design will facilitate the implementation of flow. For example, if a four-inch-square part is being made, the machines should be very close together. If they are separated by a wide distance, the tendency will be to process the parts in batch, not flow, methodology.

Determine the Takt Time

The first step when defining cell layout is to determine the customer demand rate, expressed as *Takt time*. ("Takt" is the German word for the wand the conductor uses to control the pace of an orchestra.) Takt time is expressed in the lowest measure that makes sense for a specific operation. Aircraft manufacturer Boeing might use a Takt time expressed in weeks, but Anheuser Busch might use a Takt time expressed in milliseconds. When determining Takt time, use a unit of measure that can be expressed in one or two digits. Use the following equation to determine Takt time.

$$\text{Takt time} = \frac{\text{Time available during the period}}{\text{Customer demand in the period}} \qquad (1\text{-}1)$$

For example, customers want 38,400 widgets for the next month and the line operates continuously, 16 hours per day. Twenty work days per month equals 320 hours per month—38,400 divided by 320 equals 120 widgets per hour, 120 per hour, is 2 per minute, or one every 30 seconds. The Takt time for this example is 30 seconds. If the line only runs one shift, the Takt time is cut in half, or 15 seconds.

Most processes do not operate continuously, so break times and lunch must be deducted from the available time.

Review the Assets

Once Takt time has been identified, the next step is to determine asset allocation. Suppose a small hydraulic press is required for an operation. It is capable of producing 60 pieces per minute, yet the Takt time is 30 seconds, or two pieces per minute. One-piece flow requires that the press produce only two pieces per minute. The options are to buy a press that produces at the Takt rate or use the current press at a much slower pace. If the latter option is chosen, the press operator should have additional responsibilities to consume the balance of the Takt time.

The phrase "right-sized equipment" means buying equipment that meets the customer demand rate or Takt time. For future business, every effort should be made to buy only right-sized equipment to operate in a flow cell. Difficult decisions must be made for existing operations in which equipment is overly capable.

Balance the Operations

If the Takt time has been determined, and the decision to use overly capable presses at a slower pace has been made, the next step is to balance operations. Take a look at Figure 1-7a, in which the cycle time of operator 3 exceeds the Takt time. With no changes to the balance, overtime will be needed to satisfy customer demand.

A one-piece flow cell with the cycle times from Figure 1-7a would not be effective. Operator 3 would control the pace and the output would be less than customer demand. Before moving to a cell operation, the implementer must eliminate the waste in operation 3 or determine how to assign some of operator 3's job steps to the other operators in the prospective cell.

Now, take a look at Figure 1-7b. The operations have been balanced and no overtime is needed. When balancing a line, always load the operations so that cycle time is just below Takt time. The normal result will be a final operation not fully loaded. Operator 5 has time to assist other operators, do some cleanup, etc. The ultimate goal is to reduce the cycle times of the first four operations enough so that the fifth operator can be shifted to another cell.

Determine the Cell Shape

There are four basic ways in which a cell can be designed, as shown in Figure 1-8.

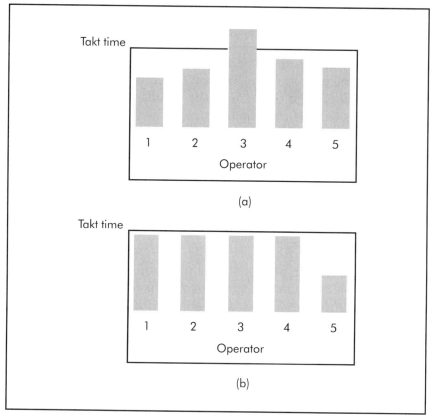

Figure 1-7. Operator cycle times.

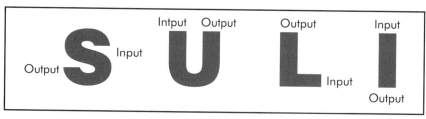

Figure 1-8. Common cell layouts.

The U-shaped cell generally uses space more effectively and reduces material handling time. S-shaped cells normally are used when the process includes a large number of operations, like an automobile assembly line. I-shaped cells are effective when deploying a small number of operations. L-shaped cells are the least effective design because space is not used efficiently. The shape of each cell will be affected by the space available in the facility.

Lay Out the Cell

For practical purposes, assume the manager has selected a U-shaped cell. Will that be a capital U or a lower case u? Cells with more operations require more space. Also, movement within a cell is limited (for example, hand-to-hand, slide, or conveyor). Conveyors are the most expensive, and hand-to-hand is not practical for large or heavy parts. Gravity slides work well with small parts, but do not physically control the flow. When developing the most efficient cell layout, keep the following in mind:

- minimize the square-footage for each operation;
- allow space for small containers of detail parts;
- keep the length of conveyors to a minimum;
- the process should flow in a counterclockwise direction (most people are right-handed), and
- operators should be located inside the U.

The final layout should look something like Figure 1-9. The first thing you notice is that the process flows from right to left. Small containers are used for detail components. The operators are located inside the U. They are positioned close together, facilitating efficient communication and minimizing the length of the conveyor. During the analysis phase, the cycle time for each operation is balanced just below the Takt time. The cell is now ready to begin making widgets.

At first, the cell runs perfectly, with each station producing one part when the following operation uses one. One-piece flow is a beautiful site to behold—so smooth, so efficient. Continuously moving conveyors move parts through the process at the Takt speed. Standardized worksheets are in place. The *visual factory*, a situation in which anyone can walk into an area and know the who, what, when, where, why, and how of the process within five minutes, is in place to identify material and process flows and normal operating parameters of the equipment.

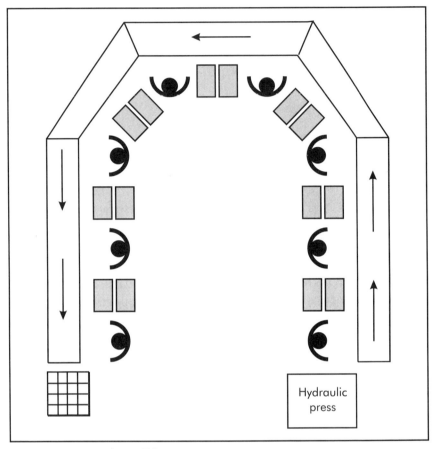

Figure 1-9. Typical U cell layout.

Suddenly, the process stops—no widgets are making it to shipping containers. What happened? The press is down; it hasn't had preventive maintenance (PM) in two years. One-piece flow will not work without effective PM. Downtime results in overtime. Make sure there is an effective PM program in place at the facility.

DEALING WITH DISCREPANCIES

Problems and discrepancies will arise during the process of establishing flow in an operation—this is natural. It is important to identify problems and their root causes. Mass manufacturing managers usually attempt to identify and solve problems in a conference room. The only way to correctly identify and solve a problem is to go to the "gemba" (the Japanese word for the operation on the floor). Managers must go where the problem exists to conduct a proper analysis.

It is critical to successful problem-solving to study the data. Opinion, gut feel, and intuition are fine outside of a manufacturing plant, but only data will point toward problem resolution. Together with the work group, the manager determines the most likely causes based on data and begins a root-cause analysis.

Once the data has been studied, develop a problem statement. Key elements to remember are that problem statements cannot include the solution, and there must be a subject and an object in the statement. Rookies typically put a solution in the problem statement, leading to a false analysis that does not unearth the root cause. A root cause has been identified when the countermeasure eliminates the problem. The following "5-Why" process depicts a good example of a root-cause analysis.

The current state: production from the door line is below customer demand rate. Overtime is expended to satisfy customer demand. A new supervisor runs the line. Downtime is excessive, and absenteeism is high.

Problem definition: door line production is behind.
Why?
Data shows machine 23 is down 12% (target is 4%).
Why?
Data shows mechanical problems account for 65% of the downtime.
Why?
Data shows the main bearing is freezing up, tripping the heat sensor.
Why?
Data shows the bearing does not stay lubricated.
Why?
Data shows that PM is not performed to schedule.

Will performing preventive maintenance on machine 23 eliminate most of the mechanical downtime, thereby increasing output and eliminating overtime? Yes. The manufacturing root cause has been identified.

INFORMATION COMMUNICATION

Hoshin kanri is the Japanese phrase for policy deployment (see Chapter 11). For a flow system to work as it should, each level in the organization must understand and support the goals of the leaders in the organization. As simple as it sounds, the "whisper syndrome" is alive and well in many facilities. Whisper a statement from person to person around a room full of people, and the ending statement will be almost completely different than the original.

Policy deployment requires that each level's goals and objectives support those of the next higher level. The number of levels in an organization can affect how closely policy deployment goals match up. More levels may mean more deviation, unless each level's goals are reviewed against those of the company leader. Effective hoshin kanri means everyone is working in the same direction.

Exceptional communication is required in operating a flow system as well. An effective *Kanban* system will result in material being delivered only in small quantities as it is needed. The Kanban is a communication device telling the supplier or supplying operation what to deliver and when to deliver it. Figure 1-10 shows a typical Kanban card. For example, Toyota receives much of its

Part Number	
XYZ123	
Using Location	Producing Location
A7-B3	**Q5-C5**
Production Quantity	Container Quantity
680	**40**

Figure 1-10. Kanban card.

material every two hours using a complex logistics system based on Kanban cards driving the just-in-time delivery system.

There are a number of ways for lean operations to communicate material requirements, both internally and externally. Radio frequency (RF) data technology has come a long way in the last few years. Sending bar code information, either from a marketplace or to a supplying department, is an effective means of communicating material requirements. There is an expense involved in the purchase of hardware and software, but it can usually be justified. If implementers choose to use an RF system, lean requires that operators control the signal to replenish. If a separate person is responsible, you've added an element of waste and increased costs.

Pull systems using Kanban cards are the most effective method of communicating material requirements. When the card is returned to the supplying operation, it delivers a signal to make a specified quantity of the required part. The system is effective for both internal and external suppliers. Some companies fax copies of the cards to suppliers, while others send the actual cards to replenish material. The first option requires suppliers to attach Kanbans to returning containers. When using the second option, the supplier receives the Kanbans and attaches them to the containers.

None of the flow functions described works well unless there is significant follow-up as the process is changing. As a change agent for the company, the implementer must make sure lean processes are being followed. Knowing that everyone is resistant to change to some degree, there is both the opportunity and the responsibility to help operators and others resolve problems, overcome fears, and understand change.

EXPECTED RESULTS

The results from converting to a flow system are easily measured:

- reduced dock-to-dock time;
- increased output per person;
- fewer labor resources required to make the product;

- fewer material handling resources required to move the product;
- less space required for in-process storage;
- less manufacturing area required;
- reduced cost, and
- improved profitability.

If an organization's personnel are ready to help the company achieve these results, develop a preliminary plan to convert to lean. Afterward, have someone experienced in lean help get the ball rolling to establish lean practices. Understand that the results of changing processes will not happen overnight. In most cases, a lean implementation takes at least two years. Alternatively, consider how competitive your company will be three years from now, if you're still using batch and queue manufacturing techniques.

Quality Feedback 2

By Perry Hall, Richard Manley, and Rachel Renick

The quality feedback system in a lean manufacturing environment ensures that every worker has a clear idea of the production process and his or her role in it. Workers are trained on how to solve problems that occur during the process. This chapter explores different facets of a quality feedback system as it is used in lean facilities. If a feedback system is to be complete, project managers must create an organizational structure that improves quality by providing information to the source of the problem. Using process parameters (for example, visual factory/visual control, built-in quality, and periodic checks) and quality targets that are properly communicated and understood causes problems to surface. Therefore, it is important to have a method in place for identifying and solving problems.

If quality standards are not accurately defined, any activities that follow are unlikely to result in a satisfactory outcome. The solution will not address the root cause of the problem, which means it is likely to recur. The process of creating a quality feedback system progresses from problem identification and short- and long-term countermeasures to follow-up and future actions.

DEFINE THE CUSTOMER

In a lean operation, the customer's role—both internal and external—must be clearly defined to create a strong foundation for lean development. If the final customer (external) is to receive a quality product, project managers must implement a zero-defect system. This allows the next (internal) customer in the lean process to add value to the product without dealing with a quality problem.

The internal customer is the next line operator in the plant who processes a product. There are many internal customers as the product moves from raw material to finished product. This definition also includes personnel within the facility who support and manage the manufacturing process. By recognizing internal customers, a degree of responsibility is assumed for passing on the product free of defects. Ultimate responsibility for guaranteeing quality resides within every member of the organization.

The external customer receives the goods after they are completed. This could be another supplier or sub-supplier, dealer, or final consumer. The person giving payment in exchange for a finished product or service is the final external customer. Examples of external customers in this situation are the final assembly plant that receives an engine manufactured by a powertrain plant, or a consumer who purchases an automobile from a dealer.

Total quality (the expected result) cannot be delivered to the final customer unless there is quality in all processes. Quality must be built into each process. To do so, a system must enable shop floor personnel to control and eliminate problems, which ensures quality and creates satisfied customers throughout the entire production process.

INSPECTION STANDARDS

If each internal and external customer is to be provided with a quality product, rules must be applied to ensure that each part meets known expectations or specifications. Within every lean organization, quality assurance should be the responsibility of every team member or operator. When quality standards are not being met, the process must immediately stop. This is an important part of any lean production system. Production of substandard quality is waste. The production process must include a structure to ensure the quality of each part or product as it is being processed.

Inspection standards should be set that enable in-station quality control. An audit of processes and standards should be included to help maintain the consistent effort needed in a lean environment. Most processes have quality checks designed into standardized work, so each part produced or assembled is individually verified to a quality standard. In addition, most lean facilities have an audit (inspection) team, which ensures that products adhere to specifications for form, fit, and function. To facilitate rapid problem-solving, roles and responsibilities must be clear. It is useful to map various problems and identify where they can best be controlled. Typically, control is best at the source, but it is not always feasible.

Table 2-1 is an example of an inspection standard chart that documents the types of checks used to ensure adherence to the standard. It identifies the functional groups or departments responsible for the check. The "X" represents a group or department with primary responsibility for confirming the part or product. The "O" signifies a group or department with responsibility for confirming that the process is followed.

Production systems are important because they control variations within the production process. Quality must be maintained at the point where value is added to the product. Process standard

Table 2-1. Inspection standard chart

Standard	Material Handling	Sub-assembly	Final Assembly	Quality Inspection
Receiving material	X			
Material dimension	X			O
Material storage	X			
Material usage	X			
Start-up line check		X	X	
First parts check		X	X	O
In-line checks		X	X	O
Off-line specification check			X	X
Process capability		O	O	X
Wrong or missing parts		X	X	O
Functional check		O	O	X
Final inspection			X	X

parameters are used to manage how the quality of a part or service is maintained. Table 2-2 outlines who is responsible for controlling activities in each area within a given process.

BUILT-IN QUALITY (JIDOKA)

Jidoka is a Japanese word that refers to a machine's ability to make judgments like that of a human. Sometimes this is referred to as *autonomation*. In a lean environment, it is imperative to use Jidoka to identify and stop problems from getting out of the process (eliminating waste). Managers must focus their teams' efforts on adding the amount of value to a part or service that meets customers' requirements, without worrying about quality.

The most fundamental effect of Jidoka is that it changes the nature of process management. It removes the need for an operator to watch a machine continuously, since the machine will stop automatically when a problem occurs. This enhancement to process management improves productivity. An operator is forced to deal with a problem at the source, virtually eliminating waste in rework and scrap production. For example, Jidoka can be applied to sensors in a stamping die to confirm the location of a panel. The press will stop and alert an operator if the sensor is tripped by an out-of-position panel to ensure that a part is not installed backwards. (See Figure 2-1a-d).

When Jidoka is connected to an Andon board (light signal board), it becomes a visual control that sounds an alarm and lights up an assigned location. Reasons for the Andon to alarm may include a tool change, quality check, error or defect, or if a machine has stopped for mechanical reasons. Operators can manage more than one machine at any given time by having the Andon alert operators to the needs of machines.

PERIODIC CHECKS

Periodic checks audit a product and/or process, and are used when it is not practical to check or gage every part. They are usually performed by manufacturing and/or production departments, based either on time (hourly, every two hours, etc.) or number of

Table 2-2. Manufacturing quality chart

Quality Engineering	Quality Engineering Raw Materials	Quality Operations	Production Process
Sets standards for making of product	Ensures all incoming material meets the specifications set by Quality Engineering to make the product	Audits or confirms quality of materials and methods to make the product	Ensures quality at the process to meet product-making specifications
Approves changes that affect the form, fit, or shape of the product and its components	Support to the vendors of raw and incoming materials	Helps control activity within and between processes with a defect(s)	Sets up equipment, methods, and manpower for each process
Pre-production planning and design; sets standards for making new products	Provides feedback to the vendor and leads problem-solving efforts on all incoming materials with defects		Tracks defects and problem-solves to prevent recurrence
Warranty and claims handling and feedback to the production process			

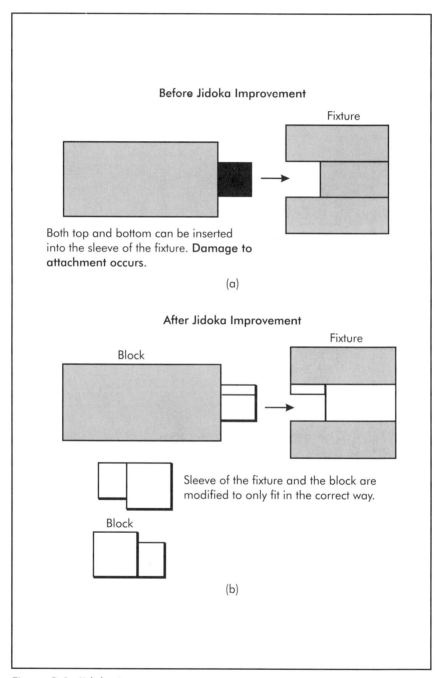

Figure 2-1. Jidoka improvements.

Quality Feedback

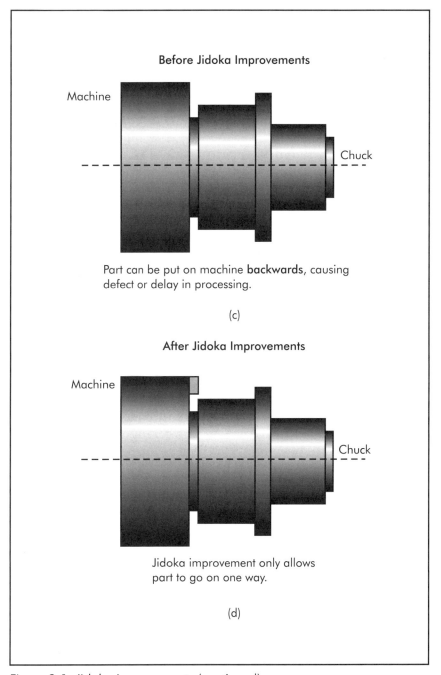

Figure 2-1. Jidoka improvements (continued).

parts produced (every 30 parts). Periodic system checks of the entire manufacturing process are generally performed by the quality department and are based on time (hourly, daily, or monthly).

Scheduled periodic product checks ensure that prescribed quality standards are met. Checks are standardized, and results are documented and posted. This document visually confirms that the check is complete and becomes a communication between shift and support groups, such as engineering and quality. Capable machines should be able to produce within the specified quality standards. More stringent checks should be used until confidence in the machine is reached. It is important to remember that this check does not replace general sorting or inspection processes. It should only be used with capable machines.

For example, a new machine or series of machines is introduced into a stable manufacturing system. After engineers qualify the ability of the machine to produce, engineering and manufacturing departments work together to stabilize the process, ensuring consistent parts production within specified quality standards. The frequency of checks during this period is greater to protect the customer. Machines should be able to stop when a defect or error occurs. Without pre-qualification and stabilization, it becomes difficult to operate using periodic check because more and more checks are necessary.

System periodic checks or audits are intended to confirm that the overall manufacturing process is working as designed. This contributes several benefits, including:

- that the product is being produced;
- manufacturing is following the agreed upon method and frequency of checks, and
- providing information to determine process capabilities.

Figure 2-2 is an example of what a manufacturing periodic check sheet looks like.

COMMUNICATION

Each department and upstream process within a facility must have a way of communicating with the preceding operator so it

Quality Feedback

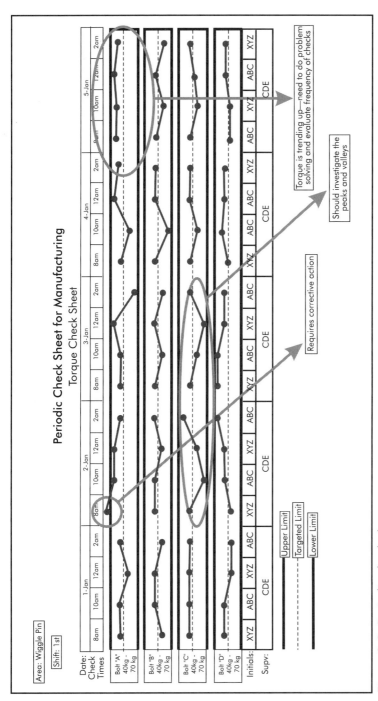

Figure 2-2. Periodic check sheet.

can strive for 100% quality. When lean implementers envision a communication process, they should keep in mind the goals of any business—to provide customers with a quality product and make a profit. Communication between operators should support these goals. The following is an example of the type of communication flow that should take place.

The Five-minute Rule

The *five-minute rule* is a communication process used to ensure that the appropriate levels of an organization are informed about quality concerns in a timely manner. The rule's precepts are that communication should be simple, direct, and structured. Team members should pass along specific, focused information quickly and concisely. Information flows quickest when the flow path eliminates ambiguous information or options. Information can be as simple as yes and no, stop and go, or okay/not okay. Branches in the flow of information should be avoided, particularly at the lower levels of the organization.

The most important aspect of the five-minute rule is not time, but the communication process (each facility's time may be longer or shorter, depending on its processes and operations). There must be a process in place to inform the proper levels of an organization, but allow the concern/problem to be dealt with at the lowest possible level.

The flow charts in Figures 2-3, 2-4, 2-5, and 2-6 are examples of what a communication plan may look like from different levels of the organization. They show the roles and responsibilities of each level in an organization. This removes any doubt as to who, what, when, and where each level of the organization must go to ensure the problem/concern is resolved.

The flow charts contain parameters that make it easy to know who and when to notify at the next level of the organization. These parameters remove any doubt about an individual's responsibility in the communication of a concern or problem. Parameters also provide instructions after the employee has notified the next level. Finally, the structure diagram provides a standard whereby improvements can be made.

Quality Feedback

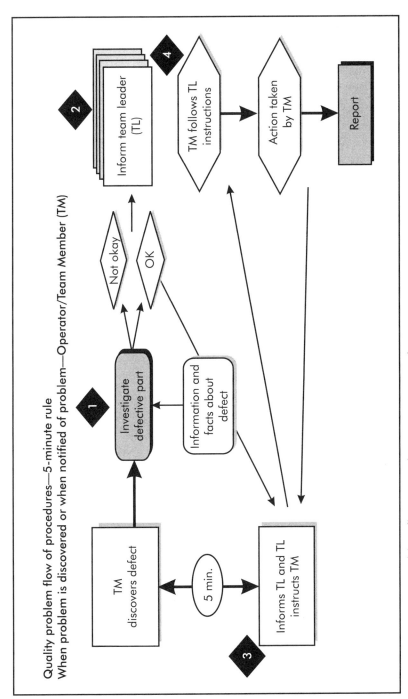

Figure 2-3. Quality problem flow procedures—5-minute rule.

Lean Manufacturing: A Plant Floor Guide

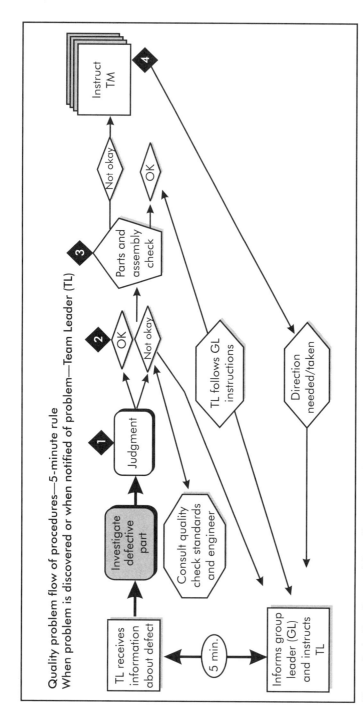

Figure 2-4. Quality problem flow procedures—5-minute rule.

Quality Feedback

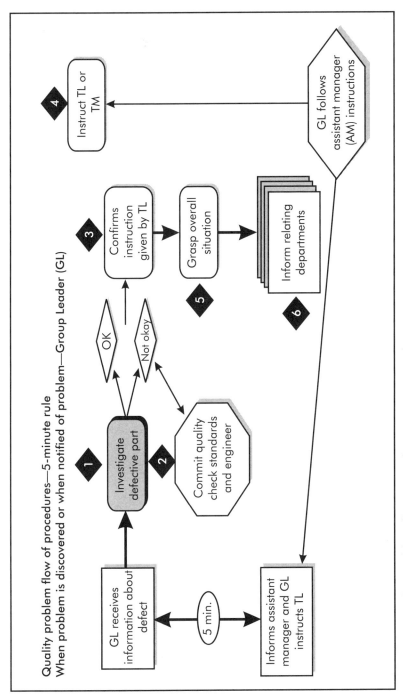

Figure 2-5. Quality problem flow procedures—5-minute rule.

Lean Manufacturing: A Plant Floor Guide

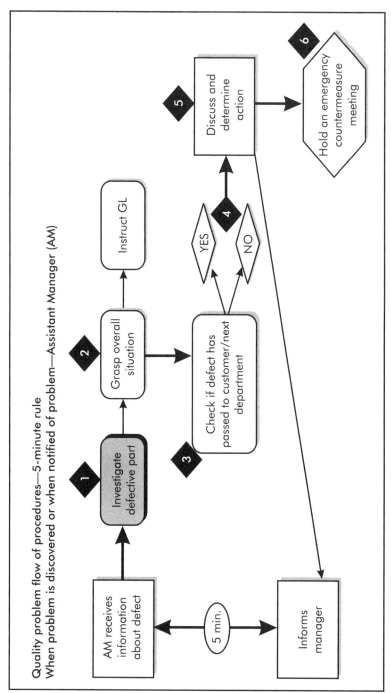

Figure 2-6. Quality problem flow procedures—5-minute rule.

ORGANIZATIONAL STRUCTURE

Quality is everyone's responsibility, and the only way to have a robust quality system is for everyone to participate. It is important to note that manufacturing and quality departments are structurally the same, but the areas of responsibility are quite different. While manufacturing is responsible for the quality of a product leaving the process, the quality department is an insurance policy to protect customers.

The quality department ensures that manufacturing is adhering to the methods and materials approved for producing the product. It also makes sure that established standards are being met, and that product functionality is never compromised. By auditing items such as frequency, correctness of process, and documentation results, the quality department can determine if a product is meeting the established standards.

All changes made to the product's assembly, as well as any changes in material, must be approved by the quality and engineering departments. If requests are made for changes in material, the quality department is the means to inform and its job is to work with vendors and suppliers. Once the changes or trials are approved, the quality department is responsible for confirming them and tracking them through the system.

PROBLEM IDENTIFICATION AND SOLVING

One of the basic principles of lean manufacturing is standardization, and that holds true in problem-solving as well. Problem-solving is best done using a systematic approach, such as the *plan-do-check-action cycle*:

- plan—define the problem, analyze for root cause, plan countermeasure implementation;
- do—implement countermeasure;
- check—confirm countermeasure implementation and effectiveness, and
- action—future activities dependent on results of check.

There are many ways to identify and prioritize problems. Problems can be picked up both reactively and proactively. The best time for problem-solving is not after a problem has occurred, but during design phases when it is still preventable. One of the most popular methods is the *Failure Mode and Effect Analysis* (FMEA). FMEA involves a cross-functional team (for example, engineers, quality personnel, shop floor team members, team leaders, etc.) brainstorming anything and everything that could go wrong in a process or with a design. Brainstorming is based on history, experience, and knowledge. After initial brainstorming is completed, the next step is to document the actions that should take place if a particular symptom occurs. The idea is to think ahead and forecast issues that might arise.

As the lean implementer begins to establish criteria for prioritizing and solving problems, he or she must prioritize the order of importance:

1. Safety or regulation items (seat belt bolt, emissions testing).
2. Functional items (motor not working).
3. Recurring appearance items (scratches on a class A surface).
4. Other minor problems (cosmetic blemishes).

These problems also can be prioritized by how far upstream the problem has traveled. The further upstream, the more important the issue.

Identify the Problem

A problem must be defined before it can be solved. Two important questions to ask are, "what is the standard?" and, "what is the actual condition?" The standard definition can come from a variety of sources (for example, written, inspection, process, standardized work, or procedures). The actual condition should be determined from witnessing the problem firsthand. Firsthand knowledge is infinitely more valuable than hearsay. Do not rely on someone else's data exclusively. Deviation from the existing standard is the difference between what should be happening and what is actually happening.

The implementer then must understand the extent of the problem. How far-reaching is it? Does it involve first or second shift, one machine or several? Does it occur at the same time every day, and the same day every week? If these factors are not known, find them out. This helps narrow the investigation.

Countermeasures

Once the problem and its extent have been identified, immediate action must be taken to eliminate it with a short-term countermeasure. This could be as simple as an extra 100% inspection to ensure the problem does not escape the inspection point. Human inspection is not 100% reliable. A better method is to install a temporary error-proofing device. A key point to remember is that the device is only temporary. Do not be fooled into thinking that 100% inspection is an effective permanent fix. A temporary fix used as a permanent solution only hides the problem and never gets to the root cause of the issue.

Acceptable countermeasures should be measurable, time-bound, and obtainable. An easy way to remember this is to "do what?", "to what?", "how much?", and "by when?".

- "Do what?" is the verb (reduce, improve, implement).
- "To what?" is the subject of the action (gap, scratches, feedback system).
- "How much?" is the measurable (5 mm, 100%).
- "By when?" is the time bound (October-end, 3/15/2001). For example, targets can be to reduce the number of scratches on the outside door handles to zero by 11/30/2001, or implement an employee program in the body shop by 1/1/2001.

Rooting Out the Cause

When managers and operators institute a countermeasure, the root cause must be determined in the process. Mistakes are often made at this juncture because of a tendency to address only the symptom, not its underlying cause. The first step is to brainstorm possible causes and analyze each one for a root cause. A good tool for root-cause analysis is the 5-Why method, which is to start with a possible cause, then ask "why?" until either the root cause is found or another problem begins to surface. When this is complete, use the "therefore" test: Start with the root cause and work backward, using the word "therefore" between phrases to see if the logic flows.

For example, during a 5S walk through, oil was noticed on the floor. Instead of just cleaning up the oil, the team decided to do root cause analysis to find out where the oil was coming from. Here's what they came up with:

Oil is on the floor.
Why is there oil on the floor?
The oil is leaking from a motor.
Why is the oil leaking from the motor?
There is a crack in the motor housing.
Why is there a crack in the motor housing?
The motor overheated.
Why did the motor overheat?
The filter was clogged.
Why was the filter clogged?
Nobody changed it.
Why didn't anyone change the filter?
It wasn't on the preventive maintenance (PM) list.

The team stopped at this point. They applied the "therefore" test:

Changing the filter wasn't on the PM list;
therefore, nobody changed the filter;
therefore, the filter was clogged;
therefore, the motor overheated;
therefore, the motor housing cracked;
therefore, oil leaked through the crack, and
therefore, there was oil on the floor.

Root-cause analysis (and problem-solving) should be conducted by asking the questions, "Why made?" and "Why shipped?" "Why made" leads to investigating and solving the reason a problem occurred. "Why shipped" leads to investigating and solving the reason a problem got to the next process. Not only did a problem occur, but for some reason it also slipped through the detection system. It is important to look at detection systems as well as the point of cause.

Permanent Solutions

Once the root cause has been found, it is time to identify permanent countermeasures. There are three types of countermeasures: short-term (which should already be in place), long-term, and recurrence prevention. Typically, there are three phases to identifying permanent countermeasures: brainstorming, trial, and final determination. During brainstorming, no ideas should be criticized. The trial phase includes a mock-up and tryout. The final determination phase includes which idea will work best, cost the least, and fully prevent the problem from happening again from the same root cause.

In the oil leak example, the short-term countermeasure identified was to clean the oil from the floor. The long-term countermeasure was to change the filter. The recurrence prevention was to add this filter (and all others like it) to the preventive maintenance system.

Once a permanent countermeasure has been identified, a plan must be developed. Everyone involved with the implementation and follow-through of the countermeasure must be included in the planning process. This encourages ownership. Ownership of an activity leads to a better success rate. The implementation plan should be detailed and include the following information: the specific activity that needs to be completed; the specific person, and the specific due date. If the plan includes all of this information, there will be no ambiguity about what needs to be done, who needs to do it, and by when it needs to be completed.

Implementation

The implementation phase involves simply getting the items on the implementation plan properly in place. For a large project, there can be daily or weekly status meetings. For smaller issues, weekly or monthly follow-ups are okay. A key point is that there is follow-up of some sort. If an item falls behind, how to get it back on schedule needs to be determined.

Part of the problem-solving process includes follow-up, which tracks the implementation status and the effectiveness of the countermeasure. These factors should be pre-determined: who will track, how the tracking will be done, and for how long. Graphs are an easy and efficient way to depict the data.

For example, suppose a problem-solving team has determined their target is to reduce scratches on the RH B-pillar paint surface to zero by 11/30/01. To measure effectiveness, the final inspector in the paint shop might keep a special log to tally the number of defects found in this area (see Table 2-3). This data is easily put in graph form (see Figure 2-7).

Table 2-3. Logging defects

Item	11/29	11/30	12/1	12/2	12/3
RH B-pillar paint scratches	III	II	0	0	II

It is easy to see now that the target has not been achieved (actual defects are greater than the target). One of two things has occurred: the countermeasure did not accurately address the root cause, or the root cause analysis was not conclusive. Either way, this indicates that the problem-solving process needs to be started over at the identifying cause stage.

CONCLUSION

The purpose of quality feedback is not fire-fighting. Its purpose is to address the root cause of problems (not just the symptoms) and prevent them from ever occurring again from the same root cause. Problems occur in any organization, of any type, and in many forms. Problems are encountered throughout the lean implementation process. The better an organization is at identifying and solving problems, the smoother the lean implementation process.

Quality Feedback

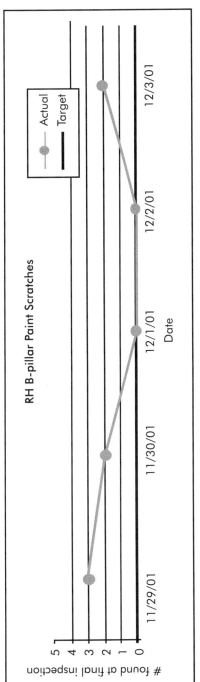

Figure 2-7. Graphing defects.

Lean Measurements and Their Use 3

By David Meier

Measurements tell everyone in an organization whether they are supporting and improving the overall performance of the team, business unit, and company. Measures are critical because they drive performance improvements. There is no value added from collecting and reporting data just to see how things are going, or to report trends.

Measures should be used primarily for collecting and analyzing data, providing immediate feedback to control problems as they occur, and sharing with the employees who can make improvements. Measurement data should be easy to collect and analyze, and enable specific action plans for continued improvement to be developed. All measures should contain specific, clearly identified target levels.

Many companies today use measurement processes (for example, quality measures, cost and/or productivity measures) to enhance the efficiency of their processes and eliminate waste. The single biggest mistake made in the measurement processes is failing to act on the information collected. Many managers have realized the importance of sharing results with all employees, but do not instruct them on how to take action based on those results. Also, they do not provide an adequate structure for solving problems that arise during production.

This chapter focuses on two areas that can be impacted during production (one during the process and one after), and the systems that support employees' ability to use the generated results. It does not focus on specific measures, but how they are used to influence the result at the two impact areas. Many different measures can be used—the one chosen depends upon the needs of the specific work area.

IMPACT AREAS

The first impact area is at the point and time of occurrence, and should invoke an immediate response. This area provides the most significant level of benefit, as issues are controlled and prevented in real time. Floor managers do not want to know at the end of the day that there were 100 defective parts. They want to know after five bad parts so that corrective action can be taken to prevent the other 95. A support system in which a production line can be stopped and assistance called (as well as using error-proofing techniques, Jidoka, and Andons) is necessary. Guidelines must cover this when appropriate, along with corrective actions that need to be taken.

The second impact area occurs after the point and time of occurrence. This method involves long-term analysis of trends and conditions, resulting in some type of corrective activity. Using the second impact area is beneficial for controlling difficult items at the time of occurrence, and leads to changes in practices, procedures, and standards.

As lean implementation proceeds, it is wise to measure progress. Several measures indicate progress in eliminating waste. For example, *dock-to-dock* (DTD) is a theoretical measure of the time it takes for material to proceed from raw materials through the facility to the shipping dock as finished goods. As flow improves and processes are aligned with pull systems, the level of in-process material decreases. Also, DTD time is reduced and the ability to turn cost to cash more quickly is improved.

During advanced implementation, the ability to more closely produce based on customer demand (the customer is the next process) by using pull-system methods will improve scheduled production. This means producing not only the correct quantity, but in the proper mix and sequence required by the customer.

ARE NEW MEASURES NEEDED?

People will perform according to the measures by which they are judged. This is a special challenge when transforming an organization from traditional to lean manufacturing. Using traditional performance measures will cause people to resist changing the habits for which they have previously been rewarded. This problem exists in a number of situations. The most common is the desire (in mass manufacturing) to produce as much as possible to improve the efficiency, or utilization, of personnel and equipment. Ideal efficiency is to make exactly what is needed, when it is needed, in the quantities required, and at the lowest cost. Realistic efficiency is to produce higher volumes so that the cost per piece is reduced; however, the overall costs are higher.

Many companies use work standards to develop consistent hours as a means of determining costs. The work area typically receives credit in the form of earned hours for the unit produced. The greater the volume produced in the time allotted, the better. However, employees may manipulate a production schedule so that products with more favorable standard rates are produced before those with less favorable rates, regardless of the scheduled or preferred sequence. (This ability to manipulate production is largely removed by using a pull system.) These decisions typically are made without regard for downstream processes; as a result, the next customer on the line is inundated with the results of all upstream manipulations. If meeting customer orders is difficult, but efficiency numbers are high, this is a likely cause.

The final process in a production line is where the ability to meet customer orders is measured. Operators there, however, have little ability to control what they receive. The end of the line, then, is the worst possible place to work because its personnel have the ultimate responsibility to meet the customer demand, but the least amount of influence.

A measurement that focuses on the final process encourages overproduction. Without a doubt, overproduction is the worst form of waste because the overproduced product has all of the previous forms of waste built into it. The goal (in a mass environment) is to maximize output without regard to whether it is needed immediately. This concept fosters the mentality that "we might as well make it because we will need it sooner or later."

Managing the Measures

If a company is going to successfully implement lean processes, it must change its measurement tools (what is being used) and its methods of deployment (how the tool is managed). Both must be changed to achieve the greatest benefit. Using measures properly encourages elevation of problems to higher levels of management within the company as warranted. It must be desirable to make problems apparent. Remember the phrase "don't shoot me, I'm just the messenger?" If the messenger is shot, the message will stop being delivered, and problems will remain suppressed and unresolved.

Effectively managing measures means that the current status or result is not as important as the trend (improving or declining). Continuous improvement is not about achieving target numbers on a daily basis—it is an ongoing cycle of improvement.

A sure-fire way to fail at lean implementation is to continue to use measures traditionally found in mass production. Performance must be measured in a way that is consistent with what lean principles require of employees. Measures must support an effort of continuous improvement.

The Need for Change: Go See the Actual Condition

The first thing to do before changing a measurement system is to determine the need for the change. The best way to do that is to "go and see" (the Japanese term "gemba" means to go to the site and see with one's own eyes). The situation must be observed firsthand and the employees who are affected must be interviewed. Do not rely on existing data or reports as a way to determine what needs to be done.

For example, are defects being passed on to the following process because productivity is so important that workers have been instructed not to stop the line for any reason? Are defects being corrected down the line or off-line, rather than at the source? Existing reports are based on current measurement methods, but they may not accurately depict what is happening.

Based on observations, determine whether problems and waste are being elevated to the appropriate managers and corrected, or whether methods to cover problems are being used. The lean project manager must be able to identify whether current practices support a lean implementation. If they do not, the current measurement tool and the methods used to apply it must be changed.

The first item to evaluate when visiting the shop floor is whether clearly defined performance standards are visible in the work area. Is the status of the work area obvious? Is an area meeting the schedule? Are employees having any difficulties? Is the work being completed correctly? These are all questions that should be answerable without asking anyone. If these questions or others regarding the performance to standards cannot be answered, there is room to improve performance measurements.

DEFINING MEASUREMENTS

Four key elements are at work in any manufacturing area—productivity, quality, safety, and cost. Each of these categories contains a number of measures to reflect related work performance (see Figure 3-1). All of the measures in each category are not relevant to all of the work areas. It is important to identify the issues that affect customers and the impact of resulting work. For example, in a captive work area (one that takes the product directly from a supplier via a conveyor), the total parts produced is not an accurate reflection of success. This work area can only produce what the supply process can give it. A more accurate measure of success here would be to calculate the amount of time the line is stopped, which the supplier incurred as a result of the customer line's inability to maintain production.

For the supply process, an important measure is how often and to what extent it shorts the customer process. Additional measures may be applicable to a particular operation. In any case, the items tracked and reported are agreed upon by the area leader and the manager. When considering productivity and quality measures, always try to think of which items are most critical from the customer and supplier processes.

Productivity

Any measure of productivity must accurately reflect an area's ability to produce what is needed, when it is needed, in the correct volume and mix, and in the most cost-efficient manner. Productivity is generally measured relative to the amount of time designated for production. For example, if the total shift time is 8

Productivity	Quality
Wait Kanban time	Customer satisfaction
Parts per labor hour	• Defects returned from following process (customer)
Total parts produced	
Line stop	Defects repaired in process
Equipment downtime (availability)	Yield
Stopping supplier process	Scrap %
Shorting customer	
Changeover time	
Cost	**Safety**
Warranty claims	Number of incidences
Scrap cost	• Injuries
Materials or supplies	• Lost work days
Total inventory	• Medical visits
Labor content	• Work-related restrictions
Raw material cost	
Other budget items	

Figure 3-1. Lean metrics.

hours, convert hours to minutes, which is 480 minutes. Subtract scheduled non-production time (lunch, breaks, meetings, cleanup time, etc.) from the minutes. This sum equals the standard operation time, or the total time the line is scheduled to operate daily. All calculations of time should use standard operation time as a base figure. Other non-scheduled time may be subtracted from this base as well, but only time during which the line is not expected to produce (causes outside its control).

Measures of Productivity

In a lean system (using Kanban or pull methods), production must be stopped to avoid the waste of overproduction. This is necessary when the customer process has stopped withdrawing product, and has not provided the necessary instruction and permission to continue producing it. The line is in *Wait Kanban* mode awaiting

Lean Measurements and Their Use

the order to resume making products required by the customer. Stopping the line due to lack of customer orders (Kanban) does not reduce the performance efficiency of the line. The amount of Wait Kanban time is recorded and subtracted from the standard operation time to get the available operation time.

It should be noted, however, that a consistently high amount of Wait Kanban time indicates an imbalance between processes, caused by greater capacity than required to meet customer demand. If this condition persists, the process Takt time should be adjusted to more closely match customer processes, and the line should be rebalanced (machinery and labor) accordingly.

Parts per labor hour (PLH) considers the variability of the workforce (which is not typical in lean), as well as the overall efficiency of the production line. For example, if there is excess line stoppage, the quantity of parts produced would not be as high, but labor content would remain the same, yielding a low parts-per-labor-hour number. This calculation is the total number of good parts produced (total parts less scrap parts), divided by the total labor hours worked (overtime hours count at the straight rate), minus any scheduled non-production time. With a fixed workforce (non-variable), the calculation is parts per hour (PPH), or the total parts produced, divided by the standard operation time (less Wait Kanban and scheduled non-production time).

$$PLH = \frac{G_T}{H_T - (K_W + N_T)} \quad (3\text{-}1)$$

where:

PLH = parts per labor hour
G_T = total good parts produced
H_T = total labor hours worked
K_W = Wait Kanban
N_T = scheduled non-production time

$$PPH = \frac{G_T}{O_T - (K_W + N_T)} \quad (3\text{-}2)$$

where:

PPH = parts per hour
O_T = standard operation time

The PPH measure is important because it shows the line's overall efficiency. It will reflect the following:

- the line is not meeting Takt time;
- excessive line stoppage, and
- high scrap rate.

This is an overall measure—to make improvements, specific reasons must be identified. If the labor content is standardized (as it should be in lean), labor costs and the amount per piece is fixed, and the preferable measure is PPH.

Total parts produced does not consider efficiency, only net production without regard to the time required. This is necessary to understand total production because all measurements, such as yield or scrap rate, are compared to the total produced and expressed as a percentage of the total.

Good parts are the total parts produced minus any scrap. Parts that can be reworked and made satisfactory are considered good parts, but are counted as rework.

Line stop time is the amount of time the line is stopped for any reason other than equipment downtime (counted separately) or lack of pull from the customer process (Wait Kanban). Causes may include parts shortages, quality issues, shortage of manpower, etc. Excessive line stop time will be reflected in a lower PPH. Specific reasons for low PPH should be recorded and analyzed if the goal is not met. Line stop generally is reported as a percentage of the standard operating time minus Wait Kanban.

Equipment downtime is the percentage of time the machine is unable to produce product during scheduled operation time. If equipment is scheduled to operate during lunch and break times, this is included in the total available time. *Availability* is the inverse percentage—a percentage of time during the total available time that the machine was available to produce. Some companies prefer to state the measure one way or the other, and either is acceptable.

Shorting the customer process and *stopping the supplier process* are measures that indicate a process' ability to produce what is needed, when it is needed, and according to the process Takt time. They should be used only if there is an ongoing problem, or if they are the best measure of process capability. These can be calculated by dividing the amount of stop time by standard operation time, minus the Wait Kanban.

It is interesting to note a related philosophy within Toyota. If a process rarely, or never, stops, it is not considered "lean enough." Many employees work very hard to limit the amount of stoppages, but if they never occur, it is likely there is too much waste (in the form of inventory or overproduction capacity). This is important to be as tight as possible without breaking. The key is to achieve a balance within the entire supply chain. This is a difficult paradigm to understand. If you perform too well, it is an indication that, in fact, you are not doing as well as possible.

A lengthy *changeover time* (CO) contributes to a lack of efficiency. Time lost is generally compensated for by running larger batch sizes (i.e., overproduction), so *quick-change* methods (sometimes referred to as Single Minute Exchange of Dies, or SMED) should be employed to reduce CO time. Progress should be recorded and tracked as improvements are made. CO time is recorded in minutes, and reflects the total time (other than breaks and lunch if work is ceased during that time) from the last good part produced before the CO, to the first good part produced after the CO is complete. The CO is not completed if a part is produced and then the equipment is stopped to confirm quality or make adjustments. Only after regular production is resumed is the CO complete.

A Word about Overall Equipment Effectiveness

A measure that has gained popularity and is considered an integral part of a Total Productive Maintenance (TPM) program is *overall equipment effectiveness* (OEE). OEE is a compilation of availability, quality rate, and performance efficiency (the ability of a machine to produce a product at the required speed). All three of these indicators are important to fully understand and improve equipment reliability, but each should be collected and recorded as separate items.

With three factors rolled into one combined measure (OEE), any improvement in one area may create a decline in another. If this happens, accurate trends are not easily seen. It is important to understand and have goals established for each element. When engaged in problem-solving and analyzing OEE, it is necessary to separate each of the three components anyway, so why not report them separately? Also, when data is grouped together (as with OEE), people have a tendency to focus on the end result and lose sight of immediate recognition, elevation, and prevention of problems as they occur. For these reasons, using composite OEE is not recommended; tracking each component separately is a wise decision.

Quality

A variety of measures are used to reflect quality performance, and an equal number of methods are used to analyze and report those results. At times they can become overly complicated, and the value of measures at the point of occurrence is lost. At the most basic level, the item produced either meets customer quality requirements (standards) or does not. Products that do not meet the standard can be either reworked to bring them within the standard or can be scrapped.

Products are categorized as either acceptable parts as produced (first run or first time through), parts requiring additional work to become acceptable (rework or repair,) or scrap parts. Products requiring minor repairs or touch-up that can be corrected by the operator within the station, and within normal Takt time cycle, are not considered repair. Examples may include minor sanding to remove flash or scratches, or additional forming or shaping.

Defects

The point of detection plays a critical role in quality improvement. The goal should be to never pass defects on to the next process (customer); certainly never to pass them to the final purchasing customer. This is a huge shift for companies that have long fostered the idea that a production line never stops, and that defects will be corrected later in the process.

Defects detected within a production area should be recorded and used within that specific process to identify and prevent the causes. Defects detected in the next sequential process should be recorded separately from those found at the original process point. In some cases, repair work is completed by the customer process and only the data collected is returned to the producing process for analysis. This data is used to understand and prevent the cause of the defect, improve processes (the in-process inspection that missed the defect), and prevent future defects from escaping. If a defect reaches the final purchasing customer, all processes in the flow assume some responsibility for not catching it.

Quality data has two primary functions as a tool for analyzing defects and installing corrective measures:

- to control and prevent the causes of defects, and
- to control and prevent the passing of defects to following processes (customers).

The primary purpose of data collection is to support the recognition, elevation, control, and correction cycle. Data calculations typically are shown as a percentage of the total product made or shipped. Customer return rates are based on the percentage of product returned from a customer relative to the total shipped to that customer. During a random quality audit or spot check, the percentage of defects detected is relative to the inspection sample size, not the total produced.

Safety

There is no question that safety of the workforce is a critical requirement for all companies. Most manufacturing companies are required to maintain statistical data of accidents and injuries to comply with Occupational Safety and Health Administration (OSHA) standards. This data and other information gathered should be used aggressively to control and prevent injuries and work-related illnesses.

Work related injuries typically fall into two broad categories:

- accidents or sudden injuries, including cuts, contusions, foreign body in the eye, etc., and
- cumulative trauma disorders (CTD) or repetitive motion injuries, including carpal tunnel syndrome, tinosynovitis (tennis elbow), thoracic outlet syndrome, etc.

Companies subject to OSHA regulations must maintain a log (OSHA 200) of all accidents and injuries, including the frequency and amount of time missed (lost work days). This log is maintained for the entire facility and is a compilation of all areas based on total hours worked.

Control and prevention of injuries and accidents is most valuable at points where they occur, which is where group-level statistics should be maintained. Repetitive motion injuries are more difficult to isolate. Dangerous conditions that may cause a trip or fall can be easily seen, whereas CTDs occur over long periods of time (sometimes years). Ergonomics (the study of the human body as it relates to the work environment) is a tool that can be used to aid in predicting future injuries. It is important that the workforce understand the importance of early reporting and diagnosis of potential problems.

The number of visits to a medical facility, or the quantity and type of work-related restrictions may indicate future problems. Using these indicators and taking corrective measures may eliminate a CTD from becoming a permanent injury. A facility's safety department should be able to help establish valuable measurements, and are a key resource for lean implementation. Local OSHA offices can be substituted if a plant does not have a safety department.

Cost

In the global economy today, fierce competition is increasing cost pressure. Sale price is determined in the market, and profits can only be increased by reducing costs. Traditional mass producers have focused primarily on the reduction of labor cost and overhead.

At Toyota (in a lean environment), labor is viewed as a fixed cost. After a person is hired, the cost of labor remains fixed. This philosophy is derived from the practice within Toyota of lifetime employment. More importantly, it is necessitated by the practice of standardizing labor content based on Takt time and standardized work. This creates a need for a fixed number of people to successfully operate the line. If a Kaizen activity results in the reduction of labor content, extra personnel could be used for other activities, such as joining a Kaizen team or new model launch team. There is no fear that a person will lose his or her job due to process improvements. This may be a confusing paradigm. At Toyota, reducing labor content is aggressively pursued, but not to eliminate staff. The purpose is to eliminate waste—extra personnel are used for other activities. The staffing level typically is about 10% lower than the ideal.

Control Waste

A large portion of the cost of producing any good or service is made up of the seven types of waste (defects, waiting, motion, over-processing, over-production, inventory, and inefficiency). Over-production, in particular, drives up costs because all forms of waste (and thus cost) are included in the extra production. Most companies have methods of reporting inventory levels, cost, and movement (turns). Existing measurements may be used in lean manufacturing, and progress can be monitored using them.

Typically, raw materials make up a large portion of today's manufacturing costs. Raw material can be wasted by over-processing (for example, using more paint than is needed or injecting more plastic material than is standard), through scrap (reject parts), or by inadequate processes (material falling on the floor, paint over-spray, or excess trim scrap).

Standards should be developed for the amount of raw material per part. Total raw material consumption should be compared to the standard quantity-per-part multiplied by the number of parts produced, which will provide a variance to standard. The calculation is as follows:

$$100 - \frac{M_P \times P_T}{M_A} \times 100 = V \quad (3\text{-}3)$$

where:

M_P = raw material variance per piece
P_T = total production
M_A = actual raw material usage
V = variance %

For example, a plastic injection molded part requires 1 lb (0.45 kg) of raw plastic to produce (total shot weight, including sprues, runners, and gates). If 10,000 parts were produced, 10,000 lb (4,536 kg) of material would be required. If the raw material consumption during this period is 12,200 lb (5,534 kg), the variance would be 18%.

In some cases, material variances can be detected immediately (as in the case of parts that can be weighed); in others, it must be analyzed as a total. In either case, the causes of variance must be investigated and controlled.

There are many other material costs that can be controlled. The cost of consumable supplies is just one area to be explored. These may include safety supplies (gloves, glasses, etc.) or production supplies that do not become part of the finished product (abrasives, tape, tools, etc.). In any case, look for the most significant waste in the area and initiate activities to collect data, analyze it, and take corrective steps to improve the situation.

THE MEASUREMENT PROCESS

When designing a measurable process, there are a number of principles that should be observed to support the objective of a lean implementation.

Basic Principles

Data should be collected as close to the point of occurrence or detection as possible. The further away that data is collected from that point, the less control there will be over the item detected.

Data should be collected manually, with pencil and paper. Automated data collection removes opportunities to immediately recognize issues and take action. One significant value of manually collecting data is that the operator is able to make an immediate and personal connection with what is happening (what they are observing). The operator also develops a sense of responsibility for elevating and controlling it. The act of manual recording makes the operator mentally and physically aware of issues as they occur. It is important not to take away the responsibility and privilege of data collection from the operator.

Data collection should be as simple as possible. Simplifying the process reduces the burden on the operator. Efforts should be made to minimize the amount of writing necessary while still maintaining the intent and content of the data collected.

Tabulating, calculating, charting, and reporting should be simplified as much as possible. A balance must be struck between functionality and appearance. Many people spend too much time and energy making beautiful graphs. A better use of time would be to correct the issues detected.

Each person, group, or department should only be responsible for a few measures. Efforts should be focused on items that most significantly impact the ability to meet goals (shown in Figure 3-1). There must be an agreement between line supervisors and upper management regarding the specific measure to use in each area. Do not try to measure all items in each category.

In addition to simplification, measures should be understandable and usable at all levels of the organization. The primary benefit will occur at the point of occurrence and recognition. At that level, there needs to be a solid understanding of how to use the measures to control and prevent issues from recurring.

Trends reported in measurements are more important than the absolute value or current result. The important thing to look for is continually improving results—there is no point at which it is good enough to remain. Likewise, if the current result is not good, it is more important that the trend is improving rather than whether the desired level is met. This philosophy is different from traditional American thinking, which is to reach a goal as fast as possible, rather than continually improving. If continuous improvement is a way of life, as goals are raised, the progress toward the new goal is a continuation of existing activities.

Improving trends in measures must equate to increasing the health of the business. Do not measure things just for the sake of measuring, to make pretty reports, or to impress the boss. Many companies can cite countless examples of data that is collected and reported but never used, or trotted out only for "dog and pony shows." Measure and improve items that will strengthen the business. Top managers should remember this when asking a subordinate for a report. Carefully evaluate its need and value beforehand.

Measures must be designed and implemented in such a way that they support lean manufacturing principles, primarily identifying and eliminating waste. Also, they must be deployed across the organization as part of company policy. Like all processes and procedures, applying measurements should be standardized. A written, documented procedure must be developed, put in place, and followed until such time that the standard is improved and changed.

THE MEASUREMENT CYCLE

The most significant opportunity for influencing and improving results is in the first measurement cycle, as issues are physically occurring. Measurements should primarily be used for immediate identification and correction of problems, as well as long-term analysis and development of permanent solutions. In addition, measurements are used as status and progress reports. Report results provide no inherent value other than raising issues to the appropriate level necessary for correction. Do not use measures strictly for reporting purposes.

Elevating Issues

Issues must be lifted to the level where resources can be gathered to resolve them. The steps in this cycle of lifting can be described as follows:

- recording,
- recognition,
- evaluation,
- elevation,
- control,
- prevention,
- continuous improvement,
- measure progress, and
- report results.

Recording

Data must be collected and recorded at the point of occurrence or detection. This data is collected in real time as the issue is detected, and should be done manually as discussed earlier in the chapter.

Recognition

The employee collecting data must be able to recognize abnormal trends. He or she must have guidelines to distinguish those that require immediate attention.

Evaluation

The purpose of evaluation is to determine an appropriate course of action. If there is a deviation from a standard or guideline, the required action is to elevate it to the next level. If an event does not cause a deviation from the standard, an operator may correct the issue and move on without elevating it. However, if there are repeated occurrences of the event that exceed the standard, the issue must be elevated. A few simple guidelines of standards will help an operator determine a course of action when evaluating an abnormal trend.

For example, an operator will determine whether a defective product is considered normal, or typical, for a particular line. The defect is recorded and corrected in a standard way. The operator then becomes aware that many of the same defects have been replicated in the past hour. Defects exceed the allotted 3% per day, so the operator must move the issue to the next level. An issue would not be elevated if it is within normal limits. Rules and directions help an operator determine whether to elevate the issue. In the case of equipment downtime, an operator may establish a set number of minutes' downtime before taking the next step to elevate the issue.

Elevation

In a lean workplace, everyone should be responsible for elevating problems. If the problem is outside of a person's control or abilities, he or she must elevate it to the next level for resolution. Everyone elevates problems, and everyone participates in the resolution. All operators have the ability, and responsibility, to stop the line and call their supervisor. Some lean tools are designed primarily to assist in elevating issues. One important device is known as an "Andon," a lighted sign and condition indicator. An audible signal also sounds to notify the supervisor of the condition.

This call system brings the supervisor to the point of recognition (both physically and mentally). Depending upon the nature of an issue, an operator may activate the supervisor call system without stopping the line. Stopping the line becomes the next stage of the process.

Control

Lean manufacturing employs a line-stop method that turns off production flow, thereby providing immediate short-term control of the issue. Stopping the line controls the spreading of the issue and elevates it to higher levels for resolution. This generally brings a lot of attention—exactly what is desired. When an issue is controlled (line stopped), it creates a real-time resolution to problems and allows operators to adhere to standards. If issues are not controlled at this level, all other activities will happen afterward—when it is too late.

Prevention

After a problem is elevated, the person responsible for answering the call observes the problem and makes an immediate assessment on how to proceed. The short-term immediate solution is to attempt to identify the source, prevent the occurrence, and resume production flow. It may not be a long-term correction, but it will work until permanent solutions can be put into place. This step completes the first cycle of the measurement process. The rest of the measurement cycle includes long-term analysis of data and continuous improvement activities.

Continuous Improvement

At the end of each shift, the data collected is tabulated and charted on a "trend" or basic line graph. Non-standard conditions can be analyzed using a tool, such as a Pareto chart, to further clarify the significance of each issue. The analysis then can be used as a basis for continuous improvement activities.

Long-term and repetitive issues are best suited for continuous improvement. Various methods can be used, including quality circles, Kaizen events, and problem-solving techniques. These activities are typically performed at the work-group level by people most likely to influence issues, but in some cases outside groups support the continuous improvement activities as well.

Measure Progress

Data can be used to measure progress toward goals by triggering immediate control and prevention, and by long-term activities to eliminate the root causes of problems.

Report Results

Measurements are used primarily as a reporting tool. The reporting process should begin a new cycle of recognition, evaluation, elevation, and control. As results are reported to the manager, he or she should recognize that there are unresolved, out-of-standard issues. An evaluation of the issue's significance can be made and elevated to the necessary level, where resources can resolve it. This process of recognition, evaluation, elevation, and control should continue up the ladder until a level is reached where the resources can be allocated to correct them.

A measurement cycle should be conducted at each point where an issue is recognized. The primary focus and benefit should be at the point of recognition, and driven back to the point of occurrence. Typically, this happens immediately at the operator level (see Figure 3-2). At the end of each work shift, the collected data is tabulated and posted on the group communication board. This brings awareness of all conditions to each member of a work group. If the issue continues and is out of standard, it is elevated to the work-group level for long-term analysis and correction. If any of the four measures are out of standard, the data should be analyzed to determine the causes (a Pareto chart may be useful).

Long-term corrective action should be initiated within the group. If an issue persists, it can be elevated outside the work group to bring attention and resources needed for a resolution. This is accomplished via a regular (usually monthly) report of the group's overall results to upper management. There are three distinct cycles of recognition through control or prevention (operator, work group, and outside the work group).

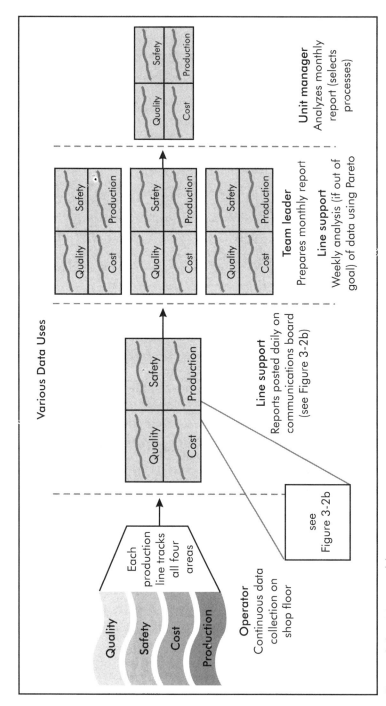

Figure 3-2a. Measurable process overview.

Hourly Requirements Based on Takt time			
1st hour 26 required ACTUAL 15	2nd hour 54 required ACTUAL 48	3rd hour 74 required ACTUAL 70	4th hour 102 required ACTUAL 92
5th hour 130 required ACTUAL 126	6th hour 150 required ACTUAL 144	7th hour 178 required ACTUAL 180	8th hour 202 required ACTUAL 211
Example report posted on communications board			

Figure 3-2b. Measurable process overview (continued).

Figure 3-3 shows a typical layout for a group communication board. Note that each of the four measures is charted using a line graph. Each graph should contain the standard (goal) line and current data. The discrepancy, or gap, is then apparent to everyone, and all members can understand the effort required to achieve the goal. Remember that a further analysis of the data (Pareto) is only necessary if an item is out of the standard. A common situation is for employees to conduct an analysis on all items, but not implement corrective actions. The analysis only provides details regarding causes, but does nothing to correct them. It is better to analyze only those items in need of activity to bring them back to standard.

Effective Tools

The cycle begins with recognition of an issue. This recognition must be transferred into a form usable by others, and is typically recorded on a data collection sheet. Collecting data is significant because abnormal trends can be detected when they are seen on the sheet. This is an important element. Avoid the use of automation or computers for data collection or tracking. Distancing occurs as operators are removed from the collection process (such as when a computer is used). Operators should connect with the process and have the ability to recognize unusual events as they occur.

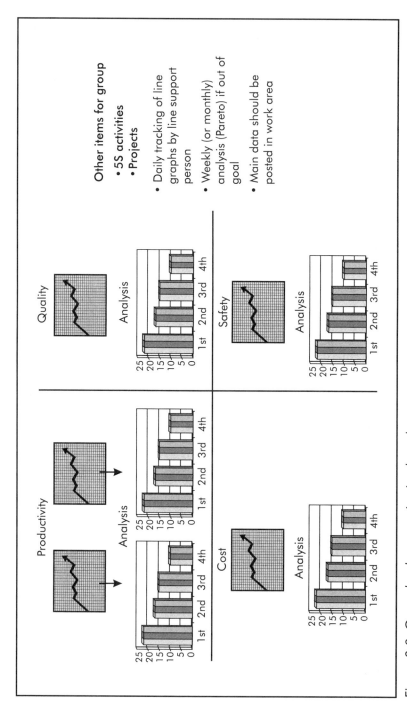

Figure 3-3. Group-level communication board.

Data collection sheets should be designed so they are easy for operators to use. After the issue is identified and the extent is understood, the data collection sheet can be designed. Consider the extent of information to collect. In most cases, too much information is collected because people want every possible issue identified. Strive to narrow issues to five top categories. For example, if equipment downtime issues are being recorded, use the top five historic causes of downtime. The leading issue may be that a product does not eject from the machine, which occurs several times each day. It is necessary to understand those items that occur frequently and repetitively.

Many items will fall into the miscellaneous category. These items happen infrequently or are one-time occurrences. For example, when tracking quality, an operator may detect a dent in the product. Dents are not typical for the process, so the operator surmises that an abnormal issue is occurring. An investigation reveals that a part of the machine is broken and is making contact with the product part, causing a dent. Since this is a random event, it is not important to classify the cause on a collection sheet. Individual classifications are used for long-term analysis and problem-solving.

Sometimes it is beneficial to use drawings or pictures to simplify the collection process or to clarify and provide additional detail. When tracking quality issues, it may be beneficial to understand where the defect occurs. Consider the amount of time that the operator would take to record each item. Try to design a sheet in such a way that only a few seconds are required for each occurrence.

Next, tabulate data at the end of a shift or, in some cases, hourly or per occurrence. Design the sheets in such a way that they can be totaled easily and transferred to tracking charts. Always remember that the process of collecting data, charting it, and reporting is all waste. There is no value-added activity in this process, so design it with an eye toward minimizing the waste. Once sheets have been designed, run through the process of tabulating and charting the data to see if it is fast and easy. Data collection sheets should have locations for the totals to be logged. The operator who collects the information typically totals all data. One hint—design all sheets and charts so that information that is transferred follows the same order and there is no need to look around the sheet for the information. Also, make all boxes and locations for information as large as possible. Use the white space on the paper.

Tabulated results are charted on a simple trend graph (line graph), with numbers written below it. It is important that graphs contain the following information in addition to the trend line:

- the standard or goal should be clearly identified;
- the chart should be titled with the name of the area and the item charted (scrap, downtime, etc.);
- show the unit of measure (pieces, minutes, percentage, etc.);
- a brief explanation of the reason if the goal was not achieved, and
- who is responsible for charting.

Use the pencil-and-paper method of charting. Computers make pretty graphs but separate people from the process. Not all employees know how to use computers and the charts must be printed every day to be updated. Use computers to create basic templates, but perform the charting by hand.

Trend charts should be posted on a communication board right in the work area (as shown in Figure 3-3). Place them in an area likely to be seen by all operators and managers. The results should be reviewed with operators at daily communication meetings, or through one-on-one meetings if necessary.

Out-of-standard Conditions

Immediately recognizing issues, elevating them, and correcting them at that time will provide the greatest benefit to controlling them. After data is collected, tabulated, and charted, it can be analyzed for out-of-standard conditions that continue to exist. When reviewing trend charts, identify items that are out of standard or where the goal has not been met on a regular basis. This should be completed on a weekly to monthly basis. If the goal has been met four out of five days, perhaps further analysis is not needed.

If a process is below its daily goal, further analysis and activity to correct the problems are required. In these cases, an analysis of the causes for out-of-standard conditions is required and the development of a Pareto-style chart or other analysis should be completed. The analysis provides specific causes for the condition and the extent (how much). This will be necessary to problem-solve. At this point, the group must begin to find a resolution or conduct PDCA (plan, do, check, act) actions to correct the issues and return to the goal.

Managers should expect to see root causes identified and action items listed for the most significant items identified in the analysis. As action items are carried out, there should be a corresponding improvement in the measurable. This information is posted on the group communication board, along with the trend charts.

On a monthly basis, charted data trends are totaled and an average for the month is determined. The monthly average is completed on an overall process trend chart and submitted to upper management for review and discussion. Areas that continue to be out of standard should be reviewed and action plans evaluated with upper management. Reports to upper managers are a consolidation of measures within a supervisor's area of responsibility. As with the entire measurables process, reported items are agreed upon by all parties. It may not be necessary to report every item because real-time data is charted in the work area. It is important for managers to visit the work site and review data at the point where they can compare what they see on charts with conditions in the work area.

Monthly reports from a department or plant may be consolidated into one document reflecting that area's overall status. Again, all items need not be reported—only the indicators that most affect the overall ability to meet customer requirements or drive up costs. It is important to not get buried in data or reports, but to reflect those items that have the most significant impact on overall results.

CONCLUSION

Today's business environment demands continuous improvement. The use of measurements as a means to drive improvement is a key element in the process of lean implementation. Most companies have measurement systems in place that do not support lean manufacturing, or are not producing the desired improvement.

Traditional measurement methods will only create traditional solutions. Instead, the cycle of recording, recognition, evaluation, elevation, control, and prevention to evaluate measurements provides lean solutions and allows them to be implemented. Each operator must know how and when to use this system. This process will require much more work and total commitment of the organization than merely implementing a data collection and reporting system, but the rewards are much more significant. A complete lean measurement system, with all of its integral parts, is a key ingredient to an overall lean manufacturing system. Its importance cannot be overlooked.

Section II:
Lean Implementation

Mapping the Value Stream 4

*By Perry Hall, Mark High,
Anthony McNaughton, and Baikuntha Sharma*

Value Stream Mapping^{SM} is a critical first step in a lean implementation because it takes a lot of the complexity and confusion out of the picture—it is based on hard facts. The mapping process involves examining and recording all of the activities that occur as a product is transformed from raw material to a finished product. Mapping creates a high-level look at total efficiency, not the independent efficiencies generated by individual cells or work groups. A value stream map contains all data relative to this flow to the customer. It captures both value added and non-value added activity.

The Value Stream Mapping process is an effective way of capturing the current situation (current state), identifying the long-term lean vision (future state), and developing a plan to get there.

CURRENT STATE MAPPING

In some respects, the journey toward lean manufacturing is no different than any other trip. The destination (vision) must be known, as well as the starting point. If there is no known starting point, it is impossible to determine whether the correct path is being traveled. One popular method to determine the starting point of a journey to lean manufacturing is the current state mapping (CSM) process. Mapping the plant floor's current condition creates a baseline from which to measure how far a facility must travel before reaching its lean destination. It also helps to point out inadequacies found in mass production. This baseline can be used to judge progress during the implementation.

CSM is a graphic depiction of what is currently happening on the floor, and it allows everyone to see and agree on what is occurring. CSM should be conducted by a cross-functional team of people. The mapping team should include representatives from manufacturing, engineering, maintenance, and production control (material handling). Data for CSM must be gathered from existing conditions on the floor, not data stored on someone's computer. To gather the information, the cross functional mapping team must walk the floor, door-to-door, following the product as it is manufactured. CSM is a pencil-and-paper process intended to get employees involved, as well as gain a better, more intimate understanding of the product, process, and information flow. Resist the urge to use a computer for this process.

CSM maps three flows for the lean journey. Product flow is the path(s) the product takes through production, before being shipped to the customer. Information flow is how information is shared and communicated during the production process. Material flow deals with how incoming material is moved and replenished, and in what quantities during production. To depict these flows, the following formation first must be gathered from the shop floor:

- run ratios—the available time divided by the number of good parts;
- scrap rates—the number of parts produced that are not salvageable;
- manpower—the number of operators in the process (actual versus required);
- work hours and schedules—the number of hours available per day, the number of shifts per day, and the number of shifts per week;
- changeover times—the amount of time it takes to change from product to product, from the last good part to the first good part;
- tool change times—the amount of time required to change tooling, from the last good part to the first good part;
- machine cycle times—the actual cycle time of each machine, from home to home, and
- inventory levels—the amount and location of all parts, including raw materials and finished goods.

This list is not all-inclusive, and there may be additional measures to record. Begin by gathering the data for one week across all shifts. Sometimes information is not available or is still being collected. It may be necessary to initiate some data collection at the processes. A representative sample must be ensured, so gather information over a suitable period that captures information on current batch sizes. In some cases, this may be more than one week. After all the information is gathered, the mapping team should begin drawing their "picture" of what the current floor condition looks like. An example of a CSM is illustrated in Figure 4-1.

To begin to understand what can be learned from the map, look at its different aspects. Begin with the customer's requirements when both drawing and reading the map. The customer's assembly plant is represented by a factory icon in the upper right-hand corner. Underneath, draw a data box recording the requirements of the customer (see Figure 4-2).

In the example, XYZ Assembly operates on two shifts and requires daily shipments. Typically, 6,000 A units, 2,000 B units, 4,000 C units, and 8,000 D units are needed every month. XYZ requests are palletized returnable trays, with 20 brackets in a tray and up to 10 trays on a pallet.

Next, draw the basic production processes (see Figure 4-3). Use process boxes. Under each processing step, write key information within the data box, such as cycle time, changeover time, and uptime.

As the flow of material is followed, project managers must find places where inventory accumulates because that is where flow stops. This is indicated with triangles to indicate "dead" flow.

Next, represent the casting supplier with another factory icon (see Figure 4-4). Use the truck icon and a broad arrow to show movement of material from the supplier to the plant.

How do the process operators know what to make and in what quantity? And what to purchase? Look at the "information flow" aspect in Figure 4-5.

The necessary materials and quantities now can be planned to meet customer demand. Figure 4-5 shows that production information is being sent to all departments, allowing them to produce as much or as little as they want. However, the departments are

Lean Manufacturing: A Plant Floor Guide

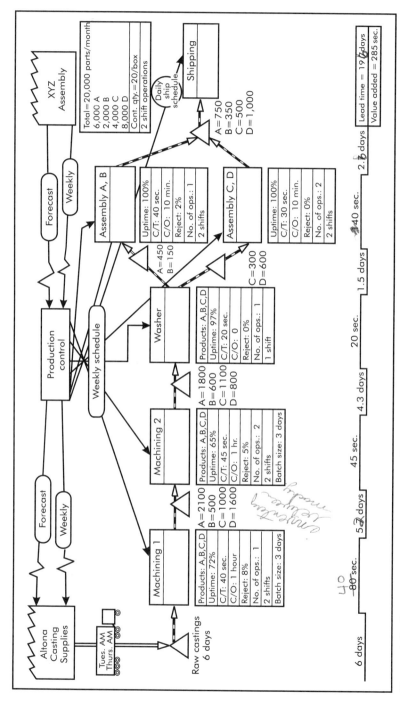

Figure 4-1. Current state map—case study example.

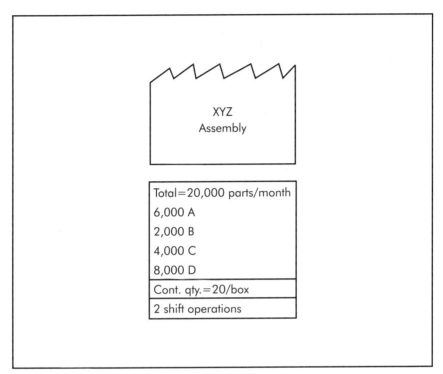

Figure 4-2. Data box recording icon.

not linked in their production number or schedules. In essence, there is no continuous flow of the material in this plant.

Now look at what the map states about production lead-time versus value-added time. Production lead-time is the amount of time it would take to produce one part through the processes, including all inventory and buffers. Processing means moving the product along to the customer. Value-added time is spent processing the product. Look at the example in Figure 4-1. In this example, production lead-time is 19.7 days, compared with only 285 seconds of value-added time.

It is now time to begin drawing a map of your current condition. Look at the symbols used in the drawing of the map (see Figure 4-6a, b). Begin by assembling a cross-functional team. This team requires training on the data to be collected, icons to be used, and the purpose of the CSM process.

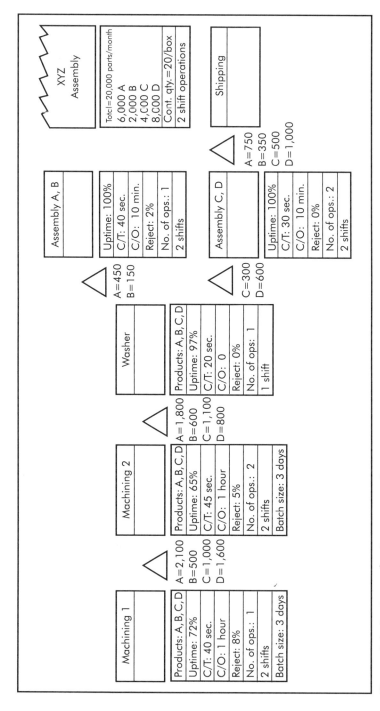

Figure 4-3. Basic production process icons.

Mapping the Value Stream

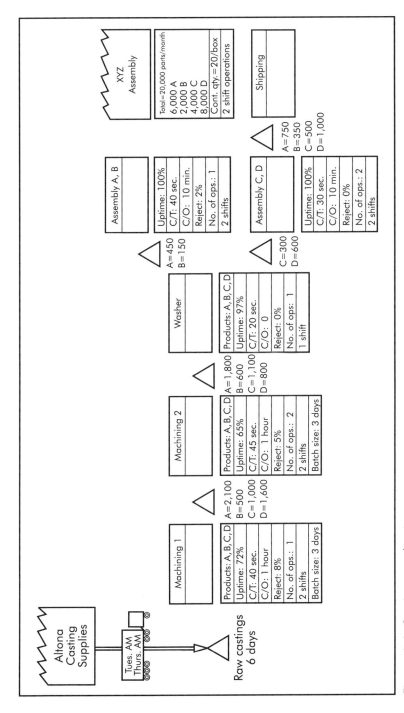

Figure 4-4. Casting supplier icon.

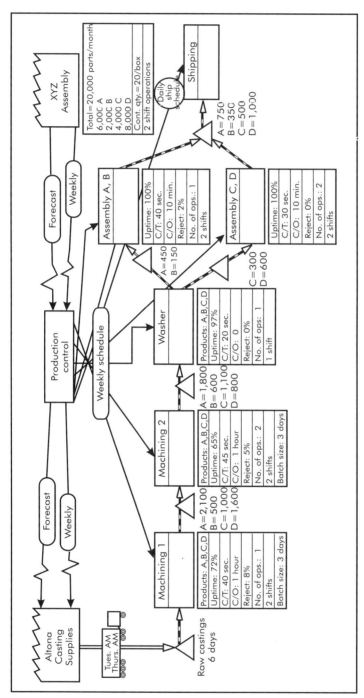

Figure 4-5. Current state information flow.

Mapping the Value Stream

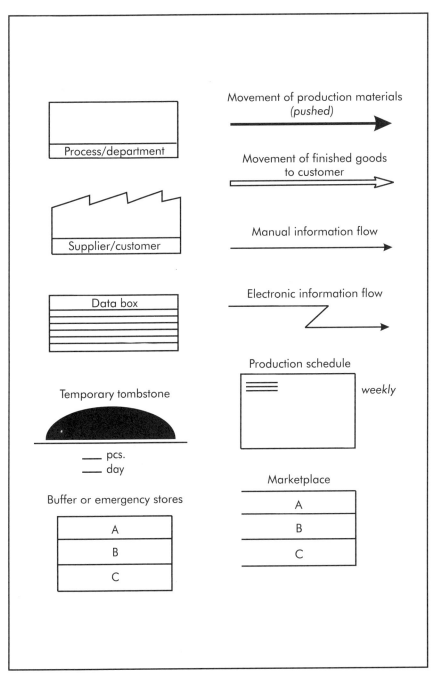

Figure 4-6a. Mapping symbols.

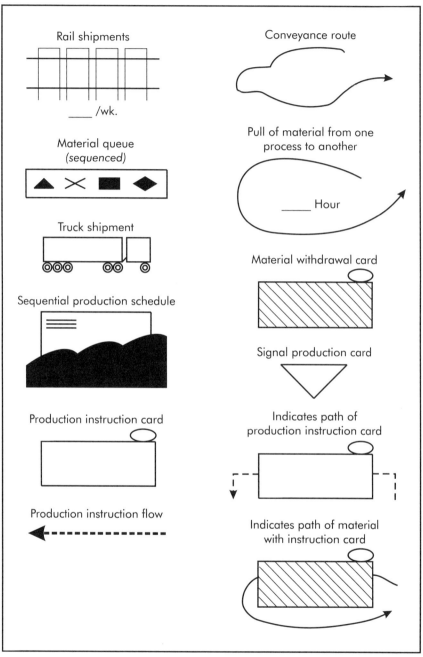

Figure 4-6b. Mapping symbols.

FUTURE STATE MAPPING

So far, this chapter has discussed ingredients that go toward creating the current state of the value stream. The CSM becomes a peg in the ground used to methodically review and build a vision, or a future state. The process of physically gathering data from the plant floor always reveals some surprises. Actual operating conditions sometimes shock stakeholders involved in the process. There is a tendency when gathering information during the CSM process to jump to solutions for the future state before the current state is completed. This can be similar to group brainstorming sessions. The same approach needs to be applied. Do not challenge any of the data, ideas, or information that comes forward. Just capture it.When the CSM is complete with product information and material flow, and its stakeholders agree that it is a reflection of the existing current status, it is time to start constructing the future state.

Analyze the Current State

The best way to gain the correct focus on what the future state should look like is to review issues from the current state. It is vital to understand the problems within the current stream from a lean viewpoint. This part of the process is critical to achieving the correct direction for an operation's future.

Many companies use the support of a sensei to provide the guidance to ensure the right decisions are achieved at this point. A *sensei* is a consultant whose mastery of lean manufacturing implementation is gained through many years of practical experience in a lean environment. Leading lean manufacturing consulting companies provide senseis to assist companies in their transition to lean. They have vast experience in making sure the appropriate data has been captured within the current state, and facilitating the establishment of the future state.

Take a closer look at the current state map in Figure 4-1. There are many symptoms of mass production. All processes in the value stream receive a schedule from production control. This is not unusual in a traditional mass environment. As a matter of fact, it would be almost impossible under current operating conditions to run the plant if the existing scheduling did not exist. What is the problem with this? It is a classic "push system." All processes (at multiple injection points) produce to schedule requirements, and are generally measured according to their ability to meet the schedule provided. When a lean practitioner discovers this condition, it is generally a precursor to other conditions that will be revealed as a result of this type of system. The various fallout from multipoint scheduling is discussed in the following sections.

Independent Efficiency

Each process within the stream is focused on its throughput from the schedule provided. If processes are held accountable and measured by this, they lose their focus on the customer's (next process) status. This traditional approach almost always strives to attain "maximum utilization."

There are three variables to consider when attempting to improve production efficiency—manpower, equipment, and materials. The traditional mass approach is to attain maximum utilization of these variables in the following order of descending priority—equipment, manpower, and materials. Under the traditional mass system, attention to manpower and materials is given only when maximum equipment usage has been attained. Attention is turned to manpower only after maximum machine use is reached. Materials usage is a poor third at best. Lean companies like Toyota have a different viewpoint. Their analysis of production costs has shown that the order of the previously mentioned production variables is incorrect, and, if anything, the order should be reversed. Toyota recognizes that emphasizing any single item without attention to the others may produce an illusion of efficiency at times. It is necessary to coordinate the three to increase efficiency. This is known at Toyota as "total efficiency."

Waste of Overproduction

In the introduction of this book, seven wastes were identified. When looking at a future state value stream, overproduction is the deadliest of all the wastes. The result of overproduction is waste in many other forms. When processes are driven by schedules, their focus on the downstream process is lost. Huge amounts of money are tied up in materials that are "sleeping" between processes. This is dead money. Reducing the amount of lost earnings by reducing in-process inventory is a source of great potential savings. This is a major aim of a lean production system.

Inability to arrange equipment and processes in accordance with a production sequence results in an awkward flow, which, in turn, leads to the following wastes:

- Work-in-process accumulates after each machine and each process.
- Material handling between processes is two or more times what it should be.
- There is poor response to changes in specifications or in products.
- There are extremely long production lead times.
- The workflow and production sequences cannot be standardized.

Waste of Correction—Quality Reject Data

A review of the current state map may reveal high reject rates through the value stream. This is not uncommon in a traditional mass environment. One problem in the current state is the amount of material "sleeping" between the processes. As a result, detection of quality issues by the next process in the value stream is delayed. This causes two problems:

1. The amount of rejects can be costly and high.
2. Identifying the root cause that allowed defects to be produced is difficult due to the length of time the defect had been produced (see Figure 4-7).

This makes effective countermeasure actions to prevent re-occurrence difficult because the waste is hidden in the inventory.

Inflexibility Within the Processes

The data gathered in the CSM process is extremely important. It allows the level of flexibility in different processes to be calibrated. Changeover time is an important issue with processes that produce a number of different parts. The current state data in

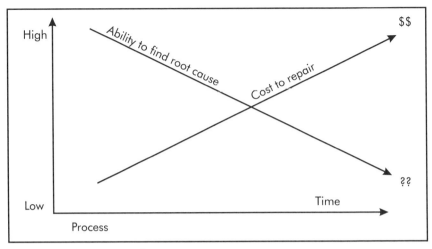

Figure 4-7. Root cause factors.

Figure 4-1 shows the machining changeover takes one hour, and the assembly process takes 10 minutes. This indicates that there is little flexibility in producing a little of all the different types throughout the hour or shift, day or week. This causes facilities to "batch build" product, which means that material sleeps between processes. It can also constrain the downstream process from producing what it needs, due to the preceding process building the wrong type of part. This inflexibility needs to be challenged for the future state.

Focus must be on the appropriate solution method when challenging issues such as changeover times. The goal is to spend the least amount of money to improve changeover times until the "method opportunities" with existing machines, materials, and manpower have been exhausted.

Process Uptime

Facilities and tooling are purchased to do a job within the operation. Any amount of unplanned downtime—breakdowns, etc.,—should not be tolerated because it is waste. CSM identifies significant amounts of unplanned downtime in the machining lines. Like the quality reject data, causes for downtime need to be identified and countermeasures implemented to provide the stability necessary for the future state.

Intrinsic Fallout

A CSM display does not always capture the total picture. It is not uncommon for the plant to experience many intrinsic problems associated with traditional mass production. Some of the issues can be:

- Shop floor "specialists" spend time throughout each day counting work-in-process and raw materials to make sure they can meet their schedule. This is a people-dependent activity, not a process-dependent activity.
- Shop floor operators and supervisors are not empowered to make daily decisions because the "schedule from above" will decide that for them.
- There is a difficult level of housekeeping to sustain. The amount of waste to support the push system ties up resources that could otherwise be available to sustain the effort.

The main issues to remember when analyzing the results of a CSM in a mass production facility are:

- Inventory may be held as containment (insurance policy) as a result of historical quality issues.
- Processes may be geographically isolated, which creates waste of conveyance and inventory.
- Plant layouts are often piecemeal—the result of company growth.
- Uptime of equipment may be reduced due to insufficient preventive maintenance or lack of resource or focus on uptime.
- Machine designs and placement may have been based on access to services such as steam, power, air, etc.

CREATING THE FUTURE STATE MAP

Keep in mind the simple goal of lean manufacturing: Produce the highest quality at the lowest total cost in the shortest lead-time, with flexibility to respond to changes. When creating the future state, managers must focus on working toward this goal.

A simple manufacturing model is illustrated in Figure 4-8. Most agree that the targets for any manufacturer are to provide customers with the product they want, within specification, at the lowest internal cost, and without injury. Of course, making a profit is an important consideration as well. It doesn't matter if the product is hamburgers or automobiles—the target is the same.

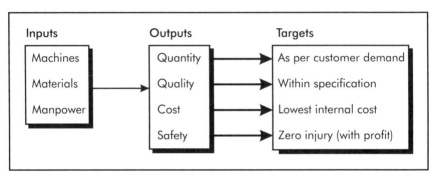

Figure 4-8. Simple manufacturing model.

The secret to successful lean companies such as Toyota is illustrated in Figure 4-9. Toyota senseis use this model to describe the choice manufacturers have within their operations.

The focus of Toyota's efforts is on which lean operational method to apply to the operation. All decisions in a lean environment are made looking toward reducing lead-time (from customer order to delivery, and from raw materials to finished product within the manufacturing plant). Lean implementers know that lead-time consists of non value-added time and value-added time. The challenge in developing the future state is to produce the customer's requirements (within specification) in the shortest lead-time and at the lowest cost. Tie this back to the example current state map (Figure 4-1), and the challenge becomes: How do managers reduce the lead-time from raw materials to finished product? The answer is to prevent overproduction.

The necessary theme for planning the future state is to "flow where you can, pull where you can't." Three tools are used to prevent overproduction and reduce leadtime—Takt time, Kanban,

Mapping the Value Stream

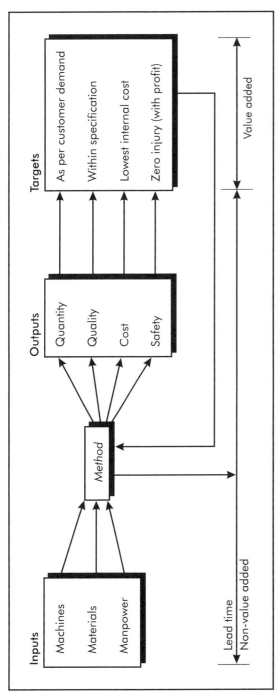

Figure 4-9. The lean manufacturing model.

and load leveling. These tools must be applied to the future state map where applicable to create an attainable vision. The effect these tools would have on the value stream is to transform the method as shown in Table 4-1.

Table 4-1. Transformation from the current to the future state

Current State	Future State
Push	Pull
Batch	Flow
Inflexible	Flexible
Top-down direction	Shop-floor managed

The future state cannot be achieved until the production value stream has the robustness necessary for the lean transition. Robustness is a high degree of reliability in equipment and machinery, resources to detect and prevent defects, and cultural stability to support the transition. If these elements are not already in place, they need to be highlighted as the future state takes shape.

A basic set of questions need to be answered to guide a process' evolution to the future state:

- What is the Takt time?
- Where can one-piece flow processing be applied?
- Are finished goods shipped to a marketplace or directly to the customer?
- At what point in the stream is it necessary to replenish the pull?
- At what point in the stream should production be scheduled?
- How and where can the product load through the stream be leveled?
- What should be the withdrawal frequency of finished goods?
- What needs to change to support the achievement of the future state?

What is the Takt Time?

Takt time is the backbone of lean operations—it is the method that synchronizes production pace throughout the value stream.

Takt time is the rate at which a process needs to produce to achieve customer demand. Sometimes it is called the customer demand rate. If products are made faster than Takt time, the result will be a build-up of work-in-process or finished goods inventory (that is, overproduction). If products are made slower than Takt time, the outcome can mean downstream shortages or the need for overtime, expedited parts, etc. This approach of operating at Takt time goes against the grain of some traditional operating methods. In some cases, it may be necessary to increase the cycle time of a process to conform to Takt time (that is, slow a process down). That way, material flow can be optimized to achieve the customer requirements "just-in-time." Material is the number one priority, then manpower, then machine.

Running to Takt time requires stability in the processes as previously mentioned. Processes need to be able to demonstrate:

- sustained uptime performance;
- first-time quality outputs, and
- flexibility to change between product types.

The Takt time formula is simple: Takt time equals available work time (net) divided by the amount of product required. If Takt time is applied to CSM for assembly processes it would look like this:

Time available = 2 shifts
= 960 minutes − (2 × 30-minute meal breaks) − (4 × 15-minute rest breaks)
= 840 minutes (50,400 seconds) available

Assembly A and B, parts required (A and B combined)

$$= \frac{8{,}000 \text{ per month}}{20 \text{ working days}} = 400 \text{ per day}$$

$$\text{Takt time} = \frac{50{,}400}{400} = 126 \text{ seconds/part}$$

Assembly C and D, parts required (C and D combined)

$$= \frac{12{,}000 \text{ per month}}{20 \text{ working days}} = 600 \text{ per day}$$

$$\text{Takt time} = \frac{50{,}400}{600} = 84 \text{ seconds/part}$$

When looking at the impact of Takt time versus current cycle times, a picture begins to emerge (see Figure 4-10).

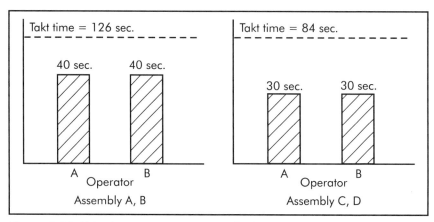

Figure 4-10. Takt time versus current cycle times.

The Takt time indicates that current cycle times are much less than the Takt time. This is not unusual in many plants. In many cases, the plant will produce whatever the daily requirement is, then re-deploy manpower to another part of the plant. This is a common case of overproduction—producing goods faster than the customer requires. This causes a build-up of work-in-process and finished goods, and generally is at risk of the symptoms previously mentioned. Before this issue is resolved, ask "where can one-piece flow processing be applied?"

Where Can One-piece Flow Processing Be Applied?

As previously defined, one-piece flow processing moves the product one part at a time without allowing it to sleep between processes.

There are a couple of opportunities to achieve one-piece flow in Figure 4-1. First, look at assembly processes. Currently, processes A and B are separated from C and D. If a solution to the current flexibility constraint (changeover) can be implemented, then the future state assembly process may look like Figure 4-11.

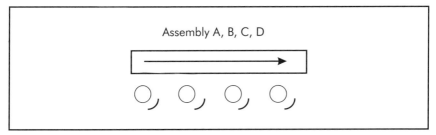

Figure 4-11. Combined assembly processes.

Revisit the Takt time calculation:

$$\frac{50{,}400 \text{ seconds available}}{1{,}000 \text{ parts daily}} = 50.4 \text{ seconds}$$

(See Figure 4-12.)

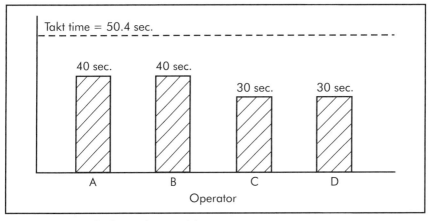

Figure 4-12. Unbalanced assembly process.

The option of balancing to Takt time now can be viewed (Figure 4-13). Note that operator D is not included. Congratulations, the first operator has been freed up to focus on more continuous improvement activities that work toward the future state.

Figure 4-13 shows re-balanced cycle times for the combined assembly processes. At this point, there are a couple of key issues worth discussing. Companies that have successfully transitioned to lean have instituted up-front policies. They have secured the

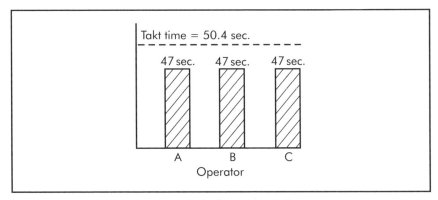

Figure 4-13. Assembly process re-balanced to Takt time.

positions of employees re-deployed because Takt time was adjusted and/or productivity improvements through waste elimination were made. The most successful of all lean companies go even further.

At other companies, the least-skilled operator within the facility is usually the one shifted when the opportunity arises. Toyota, however, has a policy that the highest skilled employee be re-deployed. They have two reasons for this—to encourage the next in line to step up as leaders, and to use the most skilled for further continuous improvement throughout the manufacturing environment. When Toyota works with its Tier 1 suppliers, it encourages this policy within the supplier's operations as well. Combining processes, as in the ongoing assembly example, is not always possible. However, project managers should always look for opportunities within the stream to do so.

Sometimes, cycle times will exceed Takt time, which constrains single-piece flow. All the options need to be reviewed, but the best one may be to reduce cycle time to less than Takt time. If this is not possible under the current operating conditions, isolate the process and hold inventory between processes to cater to the constrained process.

There is an opportunity to flow the product through the machine and washer processes. Since there is significant changeover time within the machining processes, they need to be treated as a batch process at this point. This prevents flow from moving all the way through the stream. The future state could look like Figure 4-14.

Mapping the Value Stream

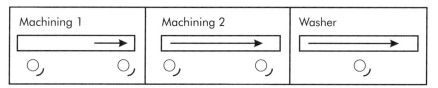

Figure 4-14. One-piece flow.

Once again, these processes need to be balanced to produce at Takt time. It is important when calculating Takt time that tool change is taken into consideration when determining the available time in the Takt formula.

If a change time of 10 minutes is targeted between products, and a batch size of one day for each product is established, then the available time would look like this:

840 − 4 × 10 = 800 minutes (48,000 seconds) of available time per day

$$\text{Takt time} = \frac{48{,}000 \text{ seconds}}{1{,}000 \text{ parts}} = 48 \text{ seconds/part}$$

There is another opportunity here. By evaluating the layout of machines and the washer, and analyzing the standardized work, Takt time may once again be re-balanced (see Figure 4-15).

Figure 4-15 shows the cycle times necessary to achieve Takt time. To achieve this type of flexibility within an operation, it is

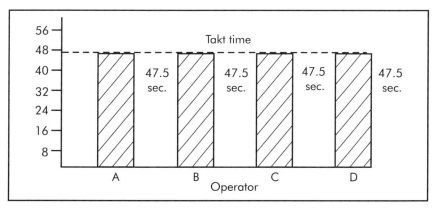

Figure 4-15. Machining/washer balanced to Takt time.

critical that operators are able to leave machines during their cycle. Cycle times are only marginally under Takt time. It is apparent that some focused cycle time reduction activity may be necessary. This provides some stability within the cycles to ensure Takt time can be met on a repeatable basis. See Figure 4-16 as an example.

Are Finished Goods Shipped to a Marketplace or Directly to the Customer?

There are several factors to be considered when determining where finished goods will be shipped:

- What is the lead-time from receiving the customer's requirements to being able to deliver that requirement?

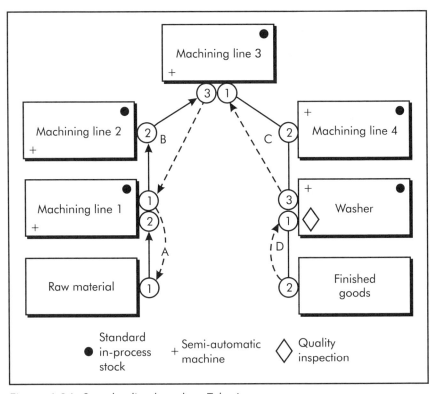

Figure 4-16. Standardized work to Takt time.

- What is the frequency of deliveries to the customer?
- Is there a reliable signal that will allow products to be built to sequence?

The most critical factor in answering these questions is the kind of withdrawal fluctuation of both volume and the mix the customer requires.

Assume for the case study (see Figure 4-1) that the customer (XYZ Assembly) is withdrawing a product range with some fluctuation of about ±20%. Also assume that the fitment point of the customer makes it difficult to deliver in sequence due to delivery lead-time. Due to this fluctuation and lead-time, produce to a finished goods marketplace. The size of the marketplace must be capable of holding enough of each product to support the withdrawal frequency, plus some safety stock to absorb fluctuations in the rate of customer withdrawal.

This should only be the starting point. Two deliveries per day to the customer immediately enables the producer to cut the size of the finished goods marketplace in half. Stability in the customer's schedule would allow the producer to build product directly for shipping to the customer, eliminating the marketplace and simply staging the product on the dock ready for shipment.

Customer withdrawal fluctuations are not uncommon. At this point, some companies rely on computer systems to predict the customer's withdrawal and produce to that prediction, but often they find themselves out of sync with the actual withdrawal. Products that are not original equipment (for example, aftermarket parts and accessories) or non-automotive hardware products, have a great range of different product types. Product types that are not consistently shipped, or products that are only manufactured when orders are placed, are best suited to be produced and shipped directly. It is not practical to hold this kind of inventory, as it would be "sleeping" in the marketplace.

At What Point in the Stream is it Necessary to Replenish the Pull?

Remember to flow where you can and pull where you can't. If the producer knows the customer is withdrawing from the finished good's marketplace, then a process needs to be established to replenish only what the customer is withdrawing. To achieve this, pull must be made from the marketplace to the preceding process, which, in this case, is the assembly process. The process prior to assembly is machining/washer. This is a batch process, so it is necessary to pull from the machining/washer marketplace. The size of this marketplace depends on the batch size of products through the machining/washer process.

When Takt time was established, the goal was to produce the requirements of the customer's range of products every day. The downstream process (assembly) is required to replenish the full range of products on a daily basis, so there must be a one-day marketplace for all products. The future state map should look like the example in Figure 4-17. By flowing where it is possible and pulling where it is not, the lead time has been reduced from 19.7 days to 5.2 days. As shown in the future state map, significant improvements are necessary throughout the stream to realize the lead-time reduction.

At What Point in the Stream Should Production be Scheduled?

The objective is to schedule as far up the stream as possible. To schedule upstream it is vital that the product flows from the scheduled point through dispatch to the customer. Schedule the stream at the assembly process if the following is in place:

- There is sufficient lead time from when the build signal is received to produce and deliver to the customer's fitment point.
- It is practical and possible to flow the product from this point forward.

In the case study (see Figure 4-17), goods are being withdrawn from the finished goods marketplace to service the customer. This condition eliminates the ability to build to sequence at the assembly process. Figure 4-17 also shows that the schedule point is at shipping. Product is being replenished as it is being withdrawn through the remainder of the stream.

Mapping the Value Stream

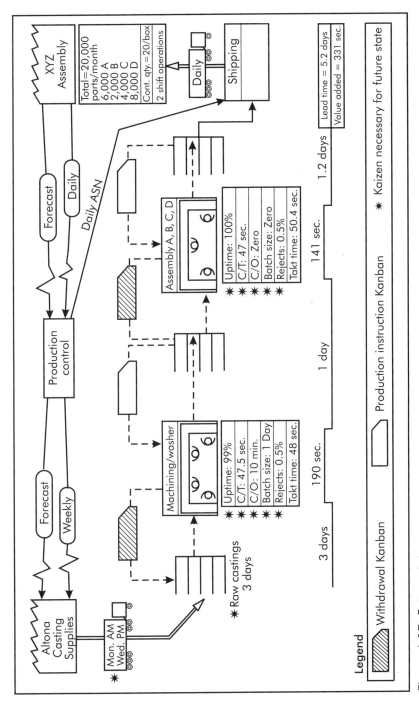

Figure 4-17. Future state map.

How and Where Can the Product Load Through the Stream be Leveled?

The constraint on leveling the product load is the flexibility of processes to change between products. Assembly, in this case, is the only process with the necessary flexibility. Machining changeover prevents load-leveling ability. In the example value stream in Figure 4-17, the customer withdraws once a day. This daily frequency creates a batch of Kanbans being returned to the assembly process for replenishment. The process provides the ability to replenish one product at a time in one batch. Most operations replenish like this to eliminate switching throughout the day.

In a lean operation, the preceding situation would not occur. The objective is to level the daily requirement of the product range across two shifts of operating time. By using this load-leveling (*heijunka*) method, the following outcomes are achieved:

- No single product is built up.
- The probability of creating a large batch of rejected product is reduced.
- There is flexibility to reduce finished goods supermarket stock levels. (If the customer increased the withdrawal frequency to greater than once per day, the facility can accommodate this without having to hold additional inventory.)
- There is an even distribution of work through the facility. Some products have greater cycle times than others, and, if produced in a batch, the facility commits the manpower to meet peak cycle times.
- There is a base to standardize the operations.
- There is an even pull of product from preceding processes in the value stream.
- The facility is protected from fluctuations in the rate of customer withdrawal.

What Should be the Withdrawal Frequency of Finished Goods?

In the example of Figure 4-17, the daily requirements of products A, B, C, and D would look like this:

A = 300/day
B = 100/day
C = 200/day
D = 400/day
Container quantity = 20

Therefore, the number of containers per day is:

A = 15
B = 5
C = 10
D = 20

The task is to spread this requirement evenly across the day.

The available time is 840 minutes, and there is a total of 50 containers to produce per day.

$$\frac{840 \text{ minutes}}{50 \text{ containers}} = 17\text{-minute intervals}$$

This is the withdrawal frequency and the pacesetter of the assembly facility. In other words, a box of parts needs to be completed every 17 minutes throughout the day to meet the customer demand. The optimum withdrawal frequency from assembly to the finished goods marketplace would be at 17-minute intervals. To withdraw more frequently would be waste. Parts would not always be there to take to the marketplace. To withdraw less frequently begins to move away from "real time management." The pulse of 17-minute intervals will highlight any issues as close to real time as practical. It becomes the pacesetter for the process.

The objective of load leveling is to allow the finished goods inventory to absorb the customer's fluctuation and protect the assembly facility from a "roller coaster" of withdrawal. To level loads, clear rules must be established for the process to be robust.

The future state map in Figure 4-18 shows the addition of the load-leveling board. It also shows the impact of the amount of finished goods necessary to support the customer's requirements if the withdrawal frequency was increased to four times per day. A further 0.8 of a day of lead-time can be cut from the stream.

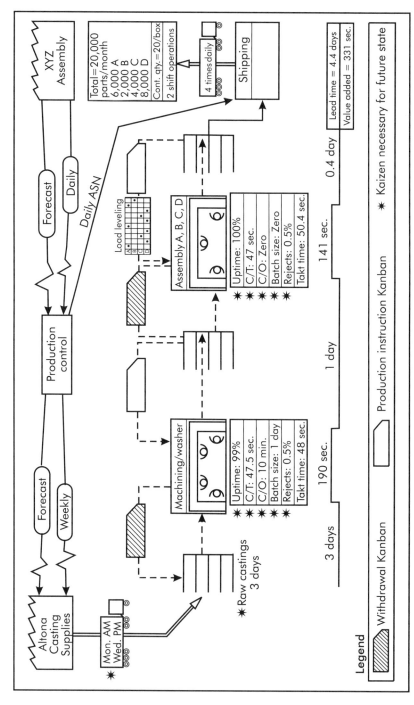

Figure 4-18. Future state map with load-leveling board and increased customer withdrawal.

What Needs to Change to Support the Achievement of the Future State?

As with any great process re-engineering initiative, there must be focus and support from all levels of the organization for a successful outcome. The transition will reveal opportunities to eliminate waste on a continuous basis. As these opportunities arise, it is paramount that they are given the necessary attention.

The future state in the case study provides many hurdles to overcome. The appropriate management and processes for issue resolution need to be in place. Carefully think through the implementation sequence before making changes. It is best to go through these basic transition steps:

1. Push—eliminate the stability and inflexibility issues.
2. Introduce pull—not lean at first.
3. When the demonstrated disciplines are in place, pull can then be made lean.

It is recommended that the pull be implemented downstream first, then worked upstream. This helps to provide stability for preceding processes.

CONCLUSION

Methodologies for Value Stream Mapping are powerful tools for providing a roadmap of the lean journey. Many companies have enjoyed great success with this approach because it highlights physical, operational, and cultural opportunities within an organization. Stability and flexibility in processes within the value stream are critical factors in achieving the vision. Mapping processes can reveal where to make changes. Remember, the future state map will remain a dream if no work is put into making it a reality.

Business Case Development 5

By David Meier

Lean implementation comes with a cost, and managers/executives must determine whether benefits outweigh costs before moving forward with a project. This chapter focuses on developing a business case. The business case is similar in scope to many methods used by companies to justify the cost of capital expenditures. It shows potential savings versus implementation costs, which yield the return on investment. This information helps the company make sound financial decisions relative to lean concepts and processes. There are a number of secondary benefits (from a savings perspective) caused by lean implementation that cannot easily be translated into dollars, such as teamwork and improved morale and attendance.

The amount of time it takes to develop the business case will vary depending on who is doing it, how much experience they have, and the time available. Typically, the business case can be developed, including the current and future state mapping, in 10 days.

Who develops the business case will vary depending on a company's resources. In many medium to large organizations, top leaders typically appoint a lean project manager and an implementation team to gather information from others as needed. Once the case is developed and finalized, the corporate CEO or president is the individual responsible for approving it. In other companies, approval may be made at the floor level. Generally, the final decision is made by the person(s) with the authority to approve expenditures and the appropriation of resources.

GETTING STARTED

"Selling the project" is a typical obstacle to implementation. Businesses today are often in fierce competition and under tight budget constraints. Before beginning, it is important to ask what goals the company wants to achieve, and how lean implementation will help achieve them. A meeting might take place with key individuals, such as the president and/or vice-president, operations or plant manager, controller, HR representatives, or implementation team members.

Determine the Stakeholders

Each situation is different, so each company must determine their own key stakeholders. In a brainstorming meeting, individuals should identify critical areas (objectives) to be improved. These may include reducing inventories (raw materials, work-in-process, and finished goods), reducing labor hours and floor space, and improving quality rates. In addition, meeting/project participants should identify secondary issues for improvement, possibly increasing the number of employee suggestions, and raising morale and employees' skill levels. A thorough list of primary and secondary improvement objectives should be generated.

Determine Goals

A difficult thing for company managers to understand is what the outcome of a lean implementation will be for the company. An important fact to remember is that the philosophy of continuous improvement means a company never arrives at a total lean state. The ultimate case is to have single-piece flow from process to process, from raw material to finished product. In reality, there are numerous situations currently making this scenario impractical; however, the ideal must be a goal to strive toward. Throughout the manufacturing process, all the components needed for that product should be available and added as needed. The objective is to continually move in the direction of single-piece flow and drive waste from processes.

How far a company goes depends on numerous factors. Single-piece flow will always be the true objective and likely remain in the foreground as a goal. Armed with the understanding that even world-class manufacturing operations contain approximately 90% waste, significant improvements in processes and measurements are within reach.

The general guidelines presented later in this chapter are considered "world class" practices for lean manufacturing. Potential savings with lean are widely recognized. A number of books currently available review potential benefits and accomplishments achieved. Renowned authorities in the area of potential benefits are James Womack and Daniel Jones. Their landmark book, *The Machine That Changed the World*, first documented significant differences between Japanese automobile manufacturers and other automakers in efficiency, quality, cost, and the ability to bring products to market quickly. Their follow-up work, *Lean Thinking*, looked beyond the automotive business and showed significant improvements in a variety of businesses. Given the authors' extensive research, their level of improvements is a guideline for other companies' potential improvement. The numbers reflected in this chapter are easily within the reach of any company converting from traditional mass production to lean manufacturing. However, the potential is dependent upon a strong effort and successful implementation.

Creating an objective for improvement is an important step in the overall success of a project. A later section in this chapter provides an illustrative range of improvement opportunities. On the low-end, savings (still very significant) are estimated after a reasonable effort and implementation success. The high end is certainly possible and has been achieved by a number of companies. The level of opportunity is dependent upon the current condition of an operation. Many companies have made some initial improvements and have "picked the low hanging fruit." As a result, they must make a more aggressive effort to reach the next level.

INTANGIBLE BENEFITS

There will be many challenges as the lean project manager and the implementation team begin building a business case. In many cases, only the items showing a verifiable cost reduction can be used to justify the project. However, a number of benefits achieved through lean yield intangible savings. Intangible items are sometimes referred to as "soft side issues." If teamwork is improved, for example, does it reduce costs? Most people agree that better teamwork is a benefit, but few agree on how much it is worth.

A substantial amount of lean's benefits is intangible. Much of the benefit does not fit onto a balance sheet. If applied properly, lean operations force a continual improvement and strengthening within the organization, and will likely impact areas for which no measures have been created to track. As a company becomes leaner, it drives the need to reduce inventories, changeover times, quality issues, and equipment failures. In essence, lean manufacturing will continue to pay dividends long after the initial implementation is complete. This provides long-term growth and strength. Lean manufacturing is much more than today's savings—it is a long-term way of doing business that provides continual payback.

For a better grasp of lean's intangible benefits, consider its core philosophy—waste elimination from all operations (removing non-value-added content). Toyota, which created the Toyota Production System (now popularly known as lean manufacturing), places its primary, continual focus on opportunities to reduce waste within all processes. This results in Toyota managers' assurances of reducing costs and strengthening the operation. Most types of waste are measured in terms of dollars saved; however, the cost of implementation, may, in fact, exceed savings projections. In most companies this would be unacceptable.

Toyota realizes that waste elimination must be pursued aggressively all the time. This realization results in a complete understanding for every member of the company that this must be carried out. There are always significant hidden benefits to be realized. Toyota understands that significant improvements can be made merely through involving employees. It is aware of a definite connection between soft-side issues and the tangible bottom-line results. For example, the company's suggestion program (soft-side) is designed to be easy for employees (and administrators) to use. They provide a large number of suggestions yielding bottom-line savings.

PLAN, PREPARE, AND DO

A lean organization has the distinct advantage of quickly and efficiently planning a process, then putting it into action immediately. Too often, many companies undergo a lengthy and difficult proposal and acceptance process, followed by a slow, controlled implementation. This is not always a bad thing; however, the cause is often a corporate culture bogged down in policies and procedures, severely restricting its ability to move. Additionally, individuals responsible for leading the charge hold up progress due to their own fears or lack of understanding on how to proceed.

It is important to maintain balance throughout the lean implementation. If all available time is spent planning, proposing, convincing, or otherwise "selling" the process, little progress will be made with the actual implementation. Implementation is where benefits are achieved, so it makes sense to focus efforts there. It is important to plan properly and gain support throughout the company, or else the implementation will be less effective. Balance here is important, and in this case a consultant may be beneficial. Consultants provide guidance regarding the speed of implementation relative to the plan. Implementation managers should spend about 20% of their time planning and preparing, and 80% implementing.

Preparing a Business Case

Managers should consider the payback for the entire plant when preparing the business case. The best approach is to look at the entire project (an implementation for the entire plant) first. This gives an overall scope and clear picture of the entire project. Sometimes it is easier to gather information for the whole plant rather than each process area. Cost information, in particular, is often looked at for an entire facility or department, rather than individual process areas. Another alternative is to implement lean in a single area as a "pilot" to prove the process and the savings. This is acceptable from an implementation standpoint; however, from a project proposal standpoint, it is limiting because the savings potential in one area generally is less significant than if a holistic approach (entire organization) is taken.

A business case should provide a comparison between the current situation and available opportunities. The amount of improvement possible is fairly consistent from one company to another. For example, reductions in inventory levels typically will be 50% or more as a company transforms from mass production to lean. This generally is true regardless of the item being produced, or other factors such as volumes and complexities. This method involves gathering data regarding total dollars available within the plant and multiplying them by a fixed percentage reduction.

As the implementer begins to build a business case, he or she will gather information on both tangible savings discussed in terms of dollars, and intangible (soft) savings that provide a benefit but are hard to capture in dollars. Tangible savings are the easiest to collect, and they are typically the savings for which businesses are looking. Inventory, labor, and floor-space reductions are the primary tangible items affected by lean transformation. Reductions in inventory and unnecessary labor required to produce a product directly impact bottom-line performance and are two of the greatest costs of operation. The potential savings are easy to determine because cost-accounting departments within the company generally monitor this information.

Table 5-1 is a guide for collecting data. Collect information that shows totals for the facility. Begin by using information collected during the current-state mapping process described in Chapter 4. The first step is to capture the current or baseline status. The worksheet explains the level of data to collect. Some costs will be an annual total (labor cost), and others will be an average of costs for a point in time (raw material inventory will be an average of the past six months to one year). Gather this material from the map and from cost accounting, production control, or other departments monitoring data. If raw data is not available, establish a method to collect it or make an educated guess. If specific information is not available, or is impractical to collect, it may be omitted. The objective is to gather the most significant items affected to show the possible returns. Fill in the appropriate information and complete the worksheet.

Table 5-1. Data collection template

Tangible Savings Potential

Category/Item	Cost/Amount	Comments
Inventory/Materials		
Raw materials		Steel, chemicals, plastic, etc., to be processed internally
Purchased components		Materials purchased from suppliers for assembly into product
Consumable goods		Products used for processing of product, such as abrasives
Work-in-process		Any material within the manufacturing process at any point
Finished goods		Saleable product ready for shipment
Carrying cost factor %		"Cost" of money tied up in inventory, usually 10–15%
Scrap cost		Total plant scrap cost per year
Rework cost		Total rework or repair cost for all processes per year
Manufacturing Labor		
Annual hours (straight time)		Total hours for previous year or 2-year average
Annual hours (overtime)		Total hours for previous year or 2-year average
Annual cost (straight time)		Total for the previous year or 2-year average
Annual cost (overtime)		Total for the previous year or 2-year average
Cost per straight-time labor hour		Includes all fringe benefits and costs
Cost per overtime/hour		Standard rate per hour without fringe times 1.5

Table 5-1. (continued)

Tangible Savings Potential

Category/Item	Cost/Amount	Comments
Equipment Maintenance		
Machine/line downtime %		Percentage of machine failure time
Equipment downtime		Cost of lost production (overtime pay, lost revenue)
Replacement parts		Purchased parts to repair or maintain equipment
Outside labor for equipment repair		Contract labor to repair or maintain equipment
Material Handling Equipment		
Annual expense for new equipment		Average total annual expense
Replacement parts		Purchased parts to repair or maintain equipment
Outside labor		Contract labor to repair equipment
Other Costs		
Value of floor space		Cost of expanding existing space or building overhead costs
Premium freight		Air freight of materials or product caused by internal failure
Warranty cost		Total cost for all warranty claims (annual)
Other costs		

Table 5-1. (continued)

Intangible Savings Potential

Category/Item	Cost/Amount	Comments
Customer Satisfaction		
Complaints or defects		Average number per 1,000 products produced
Number of past due orders		Average number per month for the last year
Past due %		Average past due orders relative to total orders
Lead time reduction		Current lead time and the value of reducing it
Delivery reliability %		Orders delivered on time, complete, and correct
Employee Involvement and Morale		
Absenteeism rate %		Unexcused absenses
Employee turnover rate %		Current turnover
Suggestions per employee		Average number of implemented suggestions per employee
Total dollar savings		Annual savings from implemented suggestions
Safety		
Lost work day incidence rate (LWDIR)		LWDIR from OSHA records
Recordable injuries		Number of recordable injuries from OSHA records
Other Items		

INVENTORY AND MATERIALS

The following few sections provide an explanation of each category and line item gathered on the data collection sheet. There may be additional opportunities in a facility. Collect the data as appropriate and evaluate the savings.

Inventory data for raw materials, purchased components, work-in-process, and consumable goods should be collected separately to get a better understanding of how the total inventory situation is divided. It is important to know the location of the most significant opportunities (and the worst problems). Initial activities should be focused in the area(s) of greatest need. If inventories are not categorized this way, use the most detailed information available.

Raw Materials

Raw materials include any material to be transformed into a product but not yet entered into the manufacturing process. They are waiting to be consumed. This includes steel, chemicals, plastics, paint, wood, paper pulp, etc. In some cases, businesses establish a consignment arrangement with suppliers, and material does not belong to (nor is it paid for by) a company until it is consumed. Lean implementers should consider this inventory, or a small percentage of it, in potential savings. Reducing the quantity stored may yield a price reduction from the supplier, because their money is tied up in inventory, which is an expense. This will be passed to customers in the form of higher costs. In some cases, suppliers store material for a company at their warehouse. This material may or may not be prepaid by your company. Consider these issues when evaluating savings potential.

Purchased Components

Purchased components come from outside suppliers just as raw materials do. Generally, these items are not transformed in a facility other than to assemble them to other products or pack in a container. Purchased components can include wire harnesses, motors, fasteners, or other components assembled into the final product. As with raw materials, consider the same situation with consignment.

Consumable Goods

Consumable goods are items used to process the product that do not become part of the finished goods. These items may include abrasives, lubricants, welding products, tools, and other miscellaneous supplies. Use of these goods should decrease as lean is implemented, as should the amount on hand.

Work-in-process

Work-in-process (WIP) is any material entering the manufacturing process. The amount of WIP in a facility is one of the largest potential savings opportunities. WIP also should be one of the easiest items to improve because each company has control of its processes and amounts. Other inventory items require cooperation from suppliers to make changes, which is more difficult. The company should have complete control of WIP. Efforts should be focused on the higher-value items in each area. Determine the average quantity of these items on hand by counting them several times. This gives a baseline comparison to monitor the level of success as the lean implementation progresses.

Finished Goods

Completed products ready for sale are considered *finished goods*. The amount of goods may be significant, depending on product variation and fluctuating demand. Many companies store high quantities of finished goods to minimize customer lead time. The cost of finished goods is high because it is the total cost of all the parts and processes. Lean implementation reduces the inventory of finished goods by using single-piece flow and smaller batch sizes. This reduces lead times and improves flexibility to quickly produce a variety of goods.

Carrying Cost

After gathering information on inventories, determine their costs. Many companies have a predetermined factor for carrying cost. Factors may include taxes, storage costs, and loss of investment potential from money invested in materials, etc. These carrying costs typically range from 8%–15%, but could easily be higher.

Scrap Cost

Scrap cost is the total cost of all materials discarded in the plant that is the plant's fault (no credit from supplier for scrap related to supplier). Record the total cost for the last fiscal year.

Rework Cost

The total cost of all repair work in the plant is called *rework*. Rework may be difficult to quantify because time spent on repair work often is not captured. Try to capture the essence of repair costs by totaling the number of people used strictly for repair, if possible. If costs cannot be associated with repair, capture the repair percentage for the plant by process area. Rework is one of the key measurables in lean, and at some point data will be collected to reflect the actual situation.

MANUFACTURING LABOR

Be wary of reducing manpower. When implemented correctly, lean manufacturing yields efficiency improvements by decreasing the number of employees needed to do the processes. An important concept to remember is that lean is a people-oriented and people-driven process. A true lean system will not be achieved without employees' support and input—implementers may improve the production method but will not acquire the true benefits of lean (continuous improvement and continued return on the investment) without employee support.

If employees are to support the process, they must have a sense of ownership in it. If they think they may lose their job as a result of their participation they surely will not participate fully. Instead, implementers should develop a plan prior to launching the process that outlines what happens to employees displaced as a result of lean. There are numerous roles employees can fill as they are displaced. For example, they may become part of a continuous improvement team, help pilot new products and processes, or take over for others on long-term leave. Lean requires a higher level of support at the floor level. These individuals may become team leaders supporting daily line operation.

When collecting labor hours, consider all hourly employees regardless of whether they are classified as direct (line workers) or indirect (supporting workers). Some opportunities to reduce the salaried workforce will arise. Due to various situations within each company, it is difficult to predict the savings. In typical situations, the salaried workforce is re-allocated. For example, many companies operate with a low number of first-line supervisors, but have many layers of management throughout. Typical lean organizations have fewer layers, but a higher ratio of supervisors to employees (management is in a support role, and each front-line supervisor can typically support about 30 people).

An initial "break-even" of labor is the target when investigating labor force reduction during a lean transition. There will not be an immediate and resulting reduction of jobs (as previously mentioned), but fewer people will be needed to perform the required tasks. Over time, the entire work force may be reduced by attrition or absorption into expanded operations. Restructure the labor allocation and create a team leader (a line-support function) role using employees removed from a line process. Lean implementation and an aggressive attack on waste in a facility reduce the number of employees required to make the product.

Do not approve the project based on the maximum labor savings shown. The potential shown is largely dependent on an organization's current structure. Examine the following areas to continue developing the business case model.

Annual Labor Hours (Straight Time)

Collect the total amount of straight-time labor hours for all hourly employees, regardless of whether they are direct or indirect workers. If the past year can be used as a representative sample, use it. If it is not, average the past few years.

Straight Time Hours with Fringe Benefits

Calculate the average cost per straight-time labor hour. This may be complicated, depending on the number of different job and pay classifications. The goal is to understand the total number of overtime hours worked, and the total cost for those hours. The cost of fringe benefits should be included in this calculation.

Annual Labor Hours (Overtime)

Reduce the overtime hours drastically, provided the same production volume is maintained. If volume increases, overtime should be lower than previous levels. One important measure is the amount of overtime hours per item produced. Collect all overtime hours currently being logged. Consider reporting by double-time or triple-time if they are significant. Otherwise, capture the total hours.

Overtime Hours (No Fringe Benefits)

The cost of fringe benefits is typically applied toward the straight-time hours, so overtime should be calculated as the base wage × 1.5 (or 2 or 3 if double- or triple-time is significant).

MANUFACTURING EQUIPMENT MAINTENANCE

The cost of equipment failure often is not apparent; however, it may cause additional overtime and it certainly has an effect on efficiency and capacity. Sometimes these costs are hidden when using traditional cost-accounting methods. A total cost structure reveals that the lack of equipment reliability directly affects the bottom line. For example, if the overall equipment failure rate is 15%, there is a 15% potential improvement in capacity, which directly reduces labor hours and overhead per item produced. All equipment breakdowns (downtime) must be considered, regardless of whether the following process is impacted, or whether the order ship date is missed.

Business Case Development

Machine or Line Downtime

Calculate the amount of time that equipment or lines are not operational when they should be. Begin with the total available time during the shift. Subtract any scheduled non-work time, such as lunch, breaks, and any other planned time the line is not running. Also, subtract time whenever the line stops because a following process has been idled, which causes the affected line to stop (the downtime is recorded at the process that is not capable of running).

Cost of Equipment Downtime

Determine the cost of lost production availability due to equipment failure. For example, what is the cost for each percentage point of downtime? An equipment failure rate of 10% equates to approximately 45 minutes per shift (assuming an eight-hour shift). In a best case scenario, an overall failure rate of 10% would cost 75% of one hour of straight-time pay with fringe benefits for affected employees. In addition, there is a cost for overhead (utilities and additional equipment expenses). Equipment reliability is a critical issue, and the true cost of unreliability must be fully understood.

Cost for Replacement Parts

Determine the total amount paid for repair parts for production-related equipment. The amount should include all parts, whether they have been installed on equipment or are spare parts. Large expenses for replacement parts may be an indication of a poor preventive maintenance program. Because equipment reliability is such an important issue, a Total Productive Maintenance (TPM) program is a must.

Outside Labor

Collect the total cost for outside labor hired to perform any maintenance, whether it be routine, preventive, or for breakdowns.

A TPM program usually results in a decrease in the cost for outside labor, particularly for repair work.

MATERIAL HANDLING EQUIPMENT

Lean manufacturing significantly changes the way material is moved within a facility. The shift is typically from a forklift moving large containers (with large quantities of parts) to tugger and dolly, and manual supply directly to the point of use. Some items will continue to be handled with forklifts, and perhaps others will be delivered directly from a truck to the line via another mechanism (for example, overhead conveyor). The bulk of items, however, should be handled manually. This methodology facilitates a single handler delivering smaller quantities of materials to larger numbers of locations. Overall, this provides for better use of manpower and equipment. Also, tugger-style equipment is less expensive than forklifts. The bulk of the forklifts can be used to load and unload trucks in the shipping and receiving operation.

New Purchases

Average the total expense for the purchase of new (or used) material handling equipment over the past three to five years. If there was no expense in the past five years, but the equipment is aged and there is an anticipated need, estimate that need. This is a potential cost avoidance.

Replacement Parts

Determine the amount paid for repair parts for material handling-related equipment. The amount should include all parts, whether they have been installed on equipment or maintained as spare parts. Large expenses for replacement parts may be an indication of a poor preventive maintenance program. Implementing a TPM program should further reduce cost.

Outside Labor for Repair or Maintenance

Determine the cost of outside labor hired to perform maintenance on all material handling equipment, whether it is routine, preventive, or for breakdowns. Do not include any items covered under warranty.

OTHER COSTS

There may be many additional items to include in the analysis. Typically, a lean implementation reduces problems in all areas. An important element of lean manufacturing is applying problem-solving skills. These skills will migrate into many areas and yield many "hidden" benefits. Evaluate other items and consider the benefits to be gained if they are improved or resolved.

Value of Floor Space

Floor space is a valuable commodity in most mass-production operations. The Flat Space Syndrome (FSS) inevitably takes over. With FSS, every flat empty space is soon filled. Two major causes of limited floor space are the amount of excess inventory and the layout of process equipment to facilitate batch production. The implementation of lean almost always yields a reduction in used floor space of about 50%. Simply cramming things together does not make these gains. The gains are made by reducing the amount of waste in the processes.

What is the value of floor space? What is it worth to the company to have more space available for no additional cost? Some companies view this as an opportunity to expand their business without additional capital expense. For others, it is an opportunity to reduce reliance on outside storage facilities. In either case there is a value, and to quantify that value, many companies use cost replacement as a basis. Some choose to value floor space based on the depreciated (book) value; others have an existing value based on overhead expense. Determine the value per square foot based on the situation at your company and multiply that by 50% of the total manufacturing space.

Premium Freight Cost

Many companies must pay a premium to transport materials to a facility or finished goods to a customer. Premium (air) freight for inbound materials is often the result of ineffective material management practices. Many problems can be averted by using a Kanban system and supporting minimum quantities. Premium freight for outbound finished goods usually results from an inability to meet a specified delivery date. Lean manufacturing reduces overall lead times and improves the ability to meet customer order dates. This yields a reduction of premium freight costs.

Warranty Cost

Determine the annual cost for warranties, replacement claims, and field service calls. Warranty claim expenses generally reduce profit margins, and are a direct reflection of the quality shipped from the plant. As overall quality is improved, there should be a corresponding decrease in warranty cost.

CUSTOMER SATISFACTION

It is difficult to measure the value of customer satisfaction. Understand that there is a cost associated with dissatisfied customers, and those costs will decrease as overall customer satisfaction improves. A number of methods are used to gage customer satisfaction. The most common are listed here. If your company uses others, calculate them as well.

Complaints or Defects per 1,000 Items Produced

The number of customer complaints per 1,000 items produced is a standard measure. If a facility's performance is outstanding, this can be measured in terms of defects per one million parts produced, or parts produced per million (PPM). Facilities measuring by 1,000 items produced will see significant improvements.

Number of Past-due Orders

The number of past-due orders is the average total number of past-due orders (did not complete according to the time line) during the past year.

The staff member capturing numbers for the business case development should attain both the number and the percentage of past-due orders. Either one alone may not seem significant, but, when considered both ways, a larger opportunity may become clear. Each past-due order indicates one dissatisfied customer. One late order is a small percentage for an operation, but for the customer it equals a rate of 100%.

Lead Time Reduction

There is a connection between lead time and the amount of finished inventory on hand. Many companies have finished goods on hand so they can quickly fill customer orders. They keep this large inventory on hand to avoid the long lead times associated with the manufacture and delivery of a product. It is neither practical nor cost effective to carry inventories of all products, so the answer is to meet customer demand more quickly. In many competitive markets, customers may make their decision to purchase based on who can deliver the product first. It is difficult to capture this cost, but the effort should be made to assess it.

Delivery Reliability

Delivery reliability is more comprehensive than the past-due measure. It includes the percentage of orders delivered on time that are correct and complete (not a partial shipment). This number provides the most accurate reflection of customer satisfaction. The customer will not be happy to receive an item on time if it is not the correct item, or if it does not work. Improving this measure goes a long way toward improving customer satisfaction. Although this is a difficult area to measure, the cost may be closely linked to warranty claims, cost of replacement parts, or shipping replacement parts.

EMPLOYEE INVOLVEMENT AND MORALE

Involving employees in meaningful work and empowering them to make a contribution has a major impact on morale. Improved employee morale generally results in a reduction of absenteeism and turnover—two items directly linked to additional cost. Some companies determine the cost to hire and train a new employee. Many companies carry extra people in the work force to cover for absentees (or simply suffer reduced production rates). Lean manufacturing provides an opportunity for the work force to become involved in continuous improvement activities and develop new skills.

Absentee Rate

The absentee rate is the number of unexcused absences. It does not include vacations, holidays, or other company approved absences, such as military leave, jury duty, funeral leave, or other medical leaves of absence.

Employee Turnover Rate

A percentage of employees each year resign (not including retirements, layoffs, or terminations by the company). The percentage of these employees who leave (versus the total number of positions within the company) is the turnover rate. These are voluntary separations initiated by the employee. A high turnover rate is an indication of a serious problem(s) at some point within the organization. Many companies use temporary workers. Do not include these workers in the turnover calculation, but a high percentage of these workers indicates potential problems. A company may use temporary employees and transition them to full-time employees (hourly) after they successfully work as a temporary. As an introduction to the workplace, the new temporary workers may be given the least desirable jobs (the jobs the full-time people do not want). This is a kind of endurance test. If they pass, they may become full-time employees. Needless to say, there is a high dropout rate. This leads to higher costs for training and processing replacements.

Number of Suggestions Per Employee

If the company has an active employee suggestion program, tabulate the number of suggestions approved and implemented,

Business Case Development

and divide it by the number of employees. This number may indicate two potential problems:

- If a program exists and the number of implemented suggestions is low, it may indicate the program is too cumbersome and difficult to use. Often, the approval process is so difficult people do not even bother unless they have a "great" idea.
- A low level of implemented suggestions may mean people just do not care (low morale).

Both issues mean there is work to be done.

Dollar Savings from Implemented Suggestions

When suggestions are made, dollars may be saved. This is where the company should see a big benefit from a suggestion program. A program should conservatively return $10 for every dollar invested (a 10:1 ratio)—a return on investment that many companies feel is excellent. Suppose the average employee was paid $100 per year for suggestions, and the company has 500 employees. That equals a pay-out of $50,000 and a savings of $500,000. It pays to have a solid suggestion program.

Determine the total dollars saved as a result of implemented suggestions. This may be a calculated savings based on what was submitted. Divide the amount by the total number of employees. As a reference point, Toyota Motor Manufacturing in Kentucky typically has a savings per employee of well over $1,000 per year, and a return rate of approximately 12:1.

SAFETY

There is no doubt that accidents and injuries are a major expense both financially and emotionally. Most accident reviews conclude that accidents occur most often when a person is doing something other than his or her regular work. Incorporating standardized work practices should have a positive impact on safety performance. Companies are required to report accidents and injuries as part of Occupational Safety and Health Administration (OSHA) requirements. Reports should be readily available.

Lost Workday Incidence Rate

The lost workday incidence report measures the number of days missed by employees resulting from accidents or injuries. It is one of the primary measures used by OSHA as an indication of a plant's safety record. This number should be available from existing OSHA reports prepared by the company's safety or human resources department.

Number of Recordable Injuries

OSHA establishes criteria for the number of recordable injuries. These are injuries severe enough to go on record. The number of recordable injuries also is found on required OSHA reports.

DETERMINING POTENTIAL SAVINGS

After the data collection sheet is complete (Table 5-1), totals can be transferred to a savings calculation worksheet (see Table 5-2). If possible, recreate the worksheet in a spreadsheet software program, such as Microsoft Excel® or Lotus 123®. This step makes it simple to adjust quantities. As dollar amounts are entered on the worksheet in the "cost" column B, multiply them by the percentage of potential savings (in columns D or E). Low savings are estimated based on a moderate level of success with the implementation. This should be a minimum target (as it represents something less than the maximum potential). High savings are estimated based on a significant level of success with the implementation (Womack and Jones 1996). If a plant is not achieving this, it is for one of two reasons: the internal process champion is not driving the process and expecting these results, or the consultant is not giving proper direction. The amount of yield in each area varies, but expect to achieve a significant level of improvement in all areas.

Opportunities within a company are impossible to determine exactly, without fully understanding the current situation and comparing it to an ideal vision of lean. Usually, the level of opportunity is much greater than imagined. In reality, the most successful companies operate with approximately 90% (or higher)

Table 5-2. Savings calculation work sheet

A Category	B Cost (from Data Collection Work Sheet)	C Potential Improvement % (Plant Average)	D Potential Savings (Low)*	E Potential Savings (High)**
Raw materials inventory (total × cost factor)		50–75%	B × 50%	B × 75%
Purchased components inventory (total × cost factor)		50–75%	B × 50%	B × 75%
Consumable goods		25–50%	B × 25%	B × 50%
Work-in-process inventory (total × cost factor)		50–75%	B × 50%	B × 75%
Finished goods inventory (total × cost factor)		50–75%	B × 50%	B × 75%
Scrap		25–50%	B × 25%	B × 50%
Rework		25–50%	B × 25%	B × 50%
Annual labor (straight time)		25–50%	B × 25%	B × 50%
Annual labor (overtime)		25–50%	B × 25%	B × 50%
Equipment downtime		10–50%	B × 10%	B × 50%
Equipment replacement parts		10–15%	B × 10%	B × 15%
Outside labor for repairs		10–15%	B × 10%	B × 15%
Annual expense for material handling equipment		25–50%	B × 25%	B × 50%
Replacement parts for material handling equipment		10–15%	B × 10%	B × 15%

Table 5-2. (continued)

A Category	B Cost (from Data Collection Work Sheet)	C Potential Improvement % (Plant Average)	D Potential Savings (Low)*	E Potential Savings (High)**
Outside labor for repair of material handling equipment		10–15%	B × 10%	B × 15%
Value of floor space ($ per ft² × total ft)		25–50%	B × 25%	B × 50%
Premium freight cost		25–50%	B × 25%	B × 50%
Warranty		25–50%	B × 25%	B × 50%
Past due orders		50–75%	B × 50%	B × 75%
Low delivery reliability		25–50%	B × 25%	B × 50%
Lead-time reduction		25–50%	B × 25%	B × 50%
Low customer satisfaction		5–10%	B × 5%	B × 10%
Low employee morale (absenteeism)		5–10%	B × 5%	B × 10%
Safety (accidents and injuries)		5–10%	B × 5%	B × 10%
Total value of suggestions per employee	Number of employees	$250–500	B × $250	B × $500
Grand totals	N/A	N/A	Sum of D	Sum of E

Table 5-2. (continued)

Category	Cost (from Data Collection Work Sheet)	Potential Improvement % (Plant Average)	Potential Savings (Low)*	Potential Savings (High)**
Projected savings per phase[†]				
Phase #1 (approximately 6 months/area)		15%	Sum of D × 15%	Sum of E × 15%
Phase #2 (approximately 6 months/area)		25%	Sum of D × 25%	Sum of E × 25%
Phase #3 (approximately 9 months/area)		30%	Sum of D × 30%	Sum of E × 30%
Phase #4 (approximately 9 months/area)		30%	Sum of D × 30%	Sum of E × 30%

[†] Phase #5 requires an additional 1–2 years and should improve overall results up to an additional 20%
* Potential savings on the low side would be achievable with a modest effort (less than industry averages)
** Potential savings on the high side are the average for all plants implementing lean

Business Case Development

non-value-added activity and the future opportunities always are huge. The limiting factor is how far an organization is willing to take the process.

Use an experienced eye to help with the evaluation—it will be invaluable. It is virtually impossible to complete a business plan and determine the true potential of a lean initiative without a clear picture of what a truly lean operation is like. This is impossible without having direct experience with a lean company. Instead, use an experienced resource either as a consultant or an internal employee. Either way, ensure that the person has first-hand experience working in a lean facility (or has a minimum of five years' experience applying lean practices). He or she should be able to assist in creating a vision of the possibility.

Projecting Savings

When projecting savings for the entire implementation process, the manager may be asked to determine returns per financial period. It is difficult to predict savings in small time increments. Savings are dependent upon the extent of implementation and opportunities in each area. A typical implementation may begin in one pilot area and spread throughout the plant. The nature of implementation makes it nearly impossible to predict its potential during specific periods without a tremendous amount of up-front analysis. It is important not to spend a great deal of time trying to determine the exact savings. It is a step that wastes time that could be spent on implementation. Remember the 80/20 rule. Spend no more than 20% of available time planning and preparing, and at least 80% implementing.

After an entire facility has completed the stability phase (the first step in the five-phase implementation process) of implementation, savings are estimated at 15% of the total possible. (See Chapter 7 for details on the five-phase implementation process.) During the stability phase, a foundational structure is being laid that may not return significant savings. However, it will begin to drive problem-solving and bring stability and reliability to the process. Each implementation must unfold according to the needs of the particular area being addressed. In many cases, flow issues within the production area need to be addressed first. In others, quality may demand immediate attention. Perhaps overall output is the primary need. The current-state mapping process should identify the primary objectives for each area, helping predict the time frame for savings to be realized. There are no savings to speak of during the current-state mapping process, although it is critical when developing a road map for improvement.

Different savings appear at different phases of implementation. Large levels of inventory are not likely to be removed from the production process until halfway through lean implementation. From there, anticipated savings are 25% of the total. Improvements to material flow and manpower efficiencies at this point yield reductions. The standardized work phase (the third of the five phases of implementation) allows all supporting processes to be closely tied together and balanced. This reduces resources and improves overall flow in the facility (and to outside suppliers). Many times in implementations there is an overlap instead of a discrete step-by-step flow.

BUILDING A BUSINESS CASE FOR A PILOT AREA

If project managers decide to implement lean in an initial pilot area, the same preceding information is gathered but only reflects items within the pilot area. Use the data collection sheet in Table 5-1. The objective is to gather as much information as possible without getting buried in data collection.

Developing a business case for a specific area is easier in some regards than for an entire plant. For example, capturing material reduction opportunities may be easier because every affected part can be identified. Each part and process can be reviewed individually and more accurately. Opportunities for improving labor efficiency and floor space savings can be easily determined. It is easier to predict the cost versus savings and timing when piloting a small area.

Some items are more difficult (or impossible) to allocate to one area because they share resources. For example, line support functions, such as material handling, engineering, and scheduling may be shared between several areas. Also, the highest level of improvement is not reached until all processes in the value stream have implemented lean. Generally, no single area is isolated from all other areas in the facility, so the best results are achieved once all areas are impacted.

Determining potential benefit may require calculations on a prorated basis. For example, an area contains 30% of the employees within a material handler's area of responsibility. If that area is reduced or eliminated, there is a 30% reduction in labor required (or the subsequent percentage is reduced). If a material handler is involved, the reduction could be based on the total number of parts handled. In any case, try to capture the potential benefit opportunity as much as possible. Do not spend a great deal of time chasing every nickel. In the end, a project usually provides greater benefits than anticipated because it is difficult to place a dollar figure on intangible savings.

Some elements do not apply to an individual area. Items such as customer satisfaction use data from the next operation (they are customers for this process) or data collected within the area relative to defects shipped to the following process. Use available data for warranty claims and other items if it can be associated to the project area.

COST OF IMPLEMENTATION

After potential savings are calculated, predict how much it will cost to implement the project. The actual expense depends on the scope of the overall project. There are two categories of expenses—internal and external. Few companies have the resources necessary to complete an entire implementation without external support of some type.

Internal Expenses

Internal expenses include salaries for project managers, coordinators, and others working on the implementation. A typical implementation should include at least one salaried project manager and a support team of salaried and hourly employees as appropriate (depending on the scope of the project). Maintenance and trades people may play an important role in the implementation as well, to support the development of Kaizen ideas, relocate equipment, and improve machine reliability. A significant time investment is required for training all employees. Finally, make allowances for employees to develop processes in their area (to perform 5S, work on problem-solving, etc.), and implement them. Consider the potential support that may be required (for example, engineering).

Some expenses may be either internal or external. For example, internal resources may be used for machinery relocation, or an outside contractor may be hired. An outside consultant with implementation expertise may need to be hired, or those resources may already be within the company (make sure potential personnel have at least five years direct experience with a truly lean company). Experienced people save time, money, and trouble during implementation, and they are worth the investment.

Some companies choose not to count internal expenses toward project implementation. It is important to understand that certain items are directly reflected on the balance sheet. Training and other activities may be done on overtime, which increases expenses. Investigate options for each situation. Do not automatically assume the way to accomplish the goal is to follow traditional methods. For example, as the number of operations is reduced, employees released from those processes can fill in so others can be trained during regular hours. The lean process requires a reconsideration of how to use resources most effectively.

External Expenses

Lean implementation does not require large capital expenditures, as external expenses fall into two categories: contracted labor, and materials and supplies. As previously mentioned, contracted labor expenses may be related to equipment relocation, using consultants or trainers, or for developing materials. Materials and supplies expenses include paint, tape, and materials for constructing communication boards, Kanbans, and Kanban boards, etc. In addition, expenses are incurred when creating flow racks to present material to operators, using dollies for manually transporting parts between processes, and for different styles of packaging. It is impossible to predict the extent of these expenses, as they result from employees' Kaizen activities. The nature of Kaizen makes it impossible to predict until the Kaizen is presented. These expenses should be moderate.

When implementing the pull system and leveled production phases, it may be necessary to acquire systems support. At this point, the project manager probably wants to integrate the Kanban system with existing or new data systems. The system tracks Kanban cycles, material usage, ordering, etc. Costs may be internal or external.

Table 5-3 can provide assistance when determining an implementation's cost. These costs are estimated for each project area, and for each phase of implementation described in Chapter 7. Costs vary according to the size of the area and overall complexity of the process. Overlap in the implementation of different areas may occur. The numbers shown reflect the amount of time required for each area. If two areas are being implemented concurrently, the number of required days doubles. Given the amount of required time for each area, it is possible for one team to implement concurrently in approximately three areas. As the implementation expands to the entire facility, each coordinator should have responsibility for three areas. If the coordinator is expected to monitor more than three areas, he or she should have additional support resources.

Leveled Production

Determining the cost reduction during the leveled production phase is difficult because the scope of impact reaches beyond the plant and into the supplier base. Leveled production relates to consistency. The volume (of parts produced) required is consistent hour by hour, day by day, and week by week. The requirement may be adjusted slightly from day to day, and changes to the production plan may be made within controlled limits. If there are large variations in production volume (which is typical in traditional manufacturing), both internal and external suppliers of the needed product must carry higher levels of stock to accommodate the wild and unpredictable swings in demand. In this case, they cannot maintain lean levels because of unpredictable demand.

Phase five will bring the facility and suppliers to leveled production. This process is fairly unique. As leveled production is

Business Case Development

Table 5-3. Return on investment calculation work sheet

A	B	C	D	E
Item	Number of People	Number of Days/Hours	Rate	Total Cost
Step #1 Current State and Future State Mapping (2–3 weeks)				
Lean manufacturing consultant	1–2	10–15 days		= B × C × D
Lean project coordinator	1	10–15 days		= B × C × D
Lean project team	Variable	10–15 days		= B × C × D
Total cost step #1				= sum
Total savings step #1 (there are no savings generated at this step)				$0
Return on investment				0%
Phase #1 Stability Implementation (6 months/area)				
Lean manufacturing consultant	1–2	15 days		= B × C × D
Lean project coordinator	1	45 days		= B × C × D
Lean project team	Variable	45 days		= B × C × D
Training—salaried employees	Variable	15–30 hours		= B × C × D
Training—hourly employees	Variable	15–30 hours		= B × C × D
Time for Kaizen implementation	Variable	32 hours		= B × C × D
Data analysis	1–2/area	45 hours		= B × C × D
Painting, cleaning, etc.	Variable	8 hours		= B × C × D
Maintenance/tooling support	2	50 hours		= B × C × D

Table 5-3. (continued)

A	B	C	D	E
Item	Number of People	Number of Days/Hours	Rate	Total Cost
Phase #1 Stability Implementation (6 months/area)				
Miscellaneous expenses (boards, cards, paint, tape, etc.)	N/A	N/A	N/A	$5,000
Outside expense—equipment relocation	N/A	N/A	N/A	$10,000
Other				
Total cost—phase #1				= sum
Savings (low—from savings calculation work sheet)				
Savings (high—rom savings calculation work sheet)				
Phase #2 Continuous Flow (6 months/area)				
Lean manufacturing consultant	1–2	15 days		= B × C × D
Lean project coordinator				
Lean project team	Variable	45 days		= B × C × D
Training—salaried employees	Variable	15–30 hours		= B × C × D
Training—hourly employees	Variable	15–30 hours		= B × C × D
Time for Kaizen implementation	Variable	32 hours		= B × C × D
Data analysis	1–2/area	45 hours		= B × C × D
Painting, cleaning, etc.	Variable	8 hours		= B × C × D
Maintenance/tooling support	2	50 hours		= B × C × D

Business Case Development

Table 5-3. (continued)

A	B	C	D	E
Item	Number of People	Number of Days/Hours	Rate	Total Cost
Phase #2 Continuous Flow (6 months/area)				
Miscellaneous expenses (boards, cards, paint, tape, etc.)	N/A	N/A	N/A	$5,000
Outside expense—equipment relocation	N/A	N/A	N/A	$30,000
Other				
Total cost—phase #2				= sum
Savings (low—from savings calculation work sheet)				
Savings (high—from savings calculation work sheet)				
Phase #3 Standardized Work (6 months/area)				
Lean manufacturing consultant	1–2	15 days		= B × C × D
Lean project coordinator	1	45 days		= B × C × D
Lean project team	Variable	45 days		= B × C × D
Training—salaried employees	Variable	15–30 hours		= B × C × D
Training—hourly employees	Variable	15–30 hours		= B × C × D
Time for Kaizen implementation	Variable	full time		no additional cost
Data analysis	1–2/area	45 hours		= B × C × D
Painting, cleaning, etc.	Variable	8 hours		= B × C × D
Maintenance/tooling support	2	50 hours		= B × C × D

Table 5-3. (continued)

A	B	C	D	E
Item	Number of People	Number of Days/Hours	Rate	Total Cost
Phase #3 Standardized Work (6 months/area)				
Miscellaneous expenses (boards, cards, paint, tape, etc.)	N/A	N/A	N/A	$3,000
Outside expense—equipment relocation	N/A	N/A	N/A	$30,000
Other				
Total cost—phase #3				= sum
Savings (low—from savings calculation work sheet)				
Savings (high—from savings calculation work sheet)				
Phase #4 Pull System (6 months/area)				
Lean manufacturing consultant	1–2	15 days		= B × C × D
Lean project coordinator	1	45 days		= B × C × D
Lean project team	Variable	45 days		= B × C × D
Training—salaried employees	Variable	15–30 hours		= B × C × D
Training—hourly employees	Variable	15–30 hours		= B × C × D
Time for Kaizen implementation	Variable	full time		no additional cost
Data analysis	1–2/area	45 hours		= B × C × D
Painting, cleaning, etc.	Variable	8 hours		= B × C × D
Maintenance/tooling support	2	50 hours		= B × C × D

Business Case Development

Table 5-3. (continued)

A	B	C	D	E
Item	Number of People	Number of Days/Hours	Rate	Total Cost
Phase #4 Pull System (6 months/area)				
Miscellaneous expenses (boards, cards, paint, tape, etc.)	N/A	N/A	N/A	$3,000
Systems support	N/A	N/A	N/A	$50,000
Other				
Total cost—phase #4				= sum
Savings (low—from savings calculation work sheet)				
Savings (high—from savings calculation work sheet)				
Phase #5 Leveled Production (2 years/entire plant)				
Lean manufacturing consultant	1	180 days		= B × C × D
Lean project coordinator	1	480 days		= B × C × D
Lean project team	Variable	480 days		= B × C × D
Training—salaried employees	Variable	90 hours		= B × C × D
Training—hourly employees	Variable	90 hours		= B × C × D
Time for implementation	Variable	full time		no additional cost
Data analysis	1–2/area	480 hours		= B × C × D
Painting, cleaning, etc.	Variable	60 hours		= B × C × D
Maintenance/tooling support	Variable	50 hours		= B × C × D

139

Table 5-3. (continued)

A	B	C	D	E
Item	Number of People	Number of Days/Hours	Rate	Total Cost
Phase #5 Leveled Production (2 years/entire plant)				
Miscellaneous expenses (boards, cards, paint, tape, etc.)	N/A	N/A	N/A	$3,000
Systems support	N/A	N/A	N/A	$50,000
Other				
Total cost—phase #5				= sum
Approximately 20% additional savings				
Total Project Cost Factors (2.5 years)				
External resource support—350 days				
External resource travel (estimated)				
Carts/dollies				
Miscellaneous materials (boards, cards, paint, tape, etc.)				
Relocation expense				

achieved, all internal processes and suppliers are able to reduce inventories and allocate resources much more accurately. Without leveled production, the daily requirement can fluctuate dramatically. To compensate for peaks in demand, each operation must have all required production items (people, machines, materials, etc.) that exceed levels necessary when production is leveled.

Achieving leveled production does not mean the lean effort is completed—actually, it is just beginning. At this point, the foundation is now in place to drive the process to new levels and areas.

BUSINESS CASE EXAMPLES

Table 5-4 is an example of the savings calculation worksheet for one facility. In it, there are numerous items for which a cost was not determined. The cost of inventories was calculated at 8% of the total cost. Also, floor space was estimated at only $1 per ft^2. This particular plant was very limited in floor space, and the value probably far exceeded $1. Take some time to evaluate all areas carefully. The more items that can be improved, the better the chances of getting approval for the project. Note that in the cost analysis (see Table 5-5), this company elected to absorb the cost of training and employee time for implementation activities. The result is a return on investment of between 4:1 and 7:1.

Cost Analysis (Example One)

The worksheets in Table 5-5 show the cost versus savings. The savings shown are taken from the savings calculation worksheet (Table 5-4). The cost of relocating equipment was estimated based on past history and the need to move several machines. The majority of equipment relocation expense is incurred during the continuous flow and standardized work phases explained in Chapter 7.

Savings Analysis (Example One)

In Table 5-6, more data was collected relative to savings potential. In some cases, plant statistics were collected but no cost was associated (such as safety). Even though there is no associated

Table 5-4. Total improvement anticipated during implementation of first four phases (2.5 years)

Category	Current Cost	Potential Improvement % (Plant Average)	Potential Savings (Low)*	Potential Savings (High)**
Finished goods inventory (carrying cost, value, etc.) Calculated at 8%	$497,172.00	–50% to –75%	$248,586.16	$372,879.24
Work-in-process inventory (carrying cost, value, etc.) Calculated at 8%	$58,867.00	–50% to –75%	$29,433.53	$44,150.29
Raw (static) inventory (carrying cost, value, etc.) Calculated at 8%	$1,061,760.00	–50% to –75%	$530,879.84	$796,319.76
Labor cost (straight time)	$4,484,000.00	–25% to –50%	$1,121,000.00	$2,242,000.00
Labor cost (overtime)	$1,144,000.00	–25% to –50%	$286,000.00	$572,000.00
Lead-time reduction		–50% to –75%	$0.00	$0.00
Scrap reduction (per year)	$240,000.00	–25% to –50%	$60,000.00	$120,000.00
Rework (estimate 3% labor)	$168,840.00	–25% to –50%	$42,210.00	$84,420.00
Machine/line downtime %		–10% to –50%	$0.00	$0.00
Machine/line performance efficiency		25% to 50%		
Material handling equipment		–50%	$0.00	$0.00

Table 5-4. (continued)

Category	Current Cost	Potential Improvement % (Plant Average)	Potential Savings (Low)*	Potential Savings (High)**
Floor space ($1/ft²)	$475,000.00	-25% to -50%	$118,750.00	$237,500.00
Premium freight		-25% to -50%	$0.00	$0.00
Past due orders		-50% to -75%	$0.00	$0.00
Warranty	$220,000.00	-10% to -15%	$22,000.00	$33,000.00
Employee morale (absenteeism)		-5%+	$0.00	$0.00
Safety		-10% to -20%	$0.00	$0.00
Customer satisfaction		?		
Delivery reliability		25%+	$0.00	$0.00
Grand totals	$8,349,639.00	N/A	$2,458,859.53	$4,502,269.29
Projected Savings Per Phase†				
Phase #1 (approximately 6 months)		15%	$368,828.93	$675,340.39
Phase #2 (approximately 6 months)		25%	$614,714.88	$1,125,567.32
Phase #3 (approximately 9 months)		30%	$737,657.86	$1,350,680.79
Phase #4 (approximately 9 months)		30%	$737,657.86	$1,350,680.79

Business Case Development

143

Table 5-4. (continued)

Per Unit Expense	
1998 production = 39,406	
Cost per unit = $630	
Estimated savings 10% (low)	$2,482,578.00
Estimated savings 25% (high)	$6,206,445.00

† Phase #5 requires an additional 1–2 years and improves overall results by up to an additional 20%
* Potential savings on the low side would be achievable with a modest effort (less than industry averages)
** Potential savings on the high side are the average for all plants implementing lean

Business Case Development

Table 5-5. Cost analysis example one—anticipated return on investment

Item	Number of People	Number of Days/Hours	Rate	Total Cost
Step #1 Current State and Future State Mapping				
Lean manufacturing consultant	1	10 days	$1,200	$12,000
Lean project coordinator	1	10 days	$200	$2,000
Lean project team	1	10 days	$300	$3,000
Data collection	2	32 hours	$12.93	$828
Introductory meeting	30	1 hour	$20.00	$600
Total cost step #1				$18,428
Total savings step #1				$0
Return on investment				0%
Phase #1 Stability Implementation (Plant Wide)				
Lean manufacturing consultant	1	60 days	$1,200	$72,000
Lean project coordinator	1	60 days	$200	$12,000
Lean project team	1	110 days	N/A	$35,000
Hourly coordinator	1	110 days	$103	$11,330
Training—hourly employees	500	15 hours	no cost	$0
Implementation hours (overtime)	500	32 hours	no cost	$0
Data analysis	20	110 hours	no cost	$0
Painting, cleaning, etc.	20	100 hours	no cost	$0
Maintenance/tooling support	10	50 hours	no cost	$0

Table 5-5. (continued)

Item	Number of People	Number of Days/Hours	Rate	Total Cost
Phase #1 Stability Implementation (Plant Wide)				
Miscellaneous expenses (boards, cards, paint, tape, etc.)	N/A	N/A	N/A	$7,500
Outside expense—equipment relocation	N/A	N/A	N/A	$30,000
Total cost—phase #1				$167,830
Savings (low)				$368,829
Savings (high)				$675,340
Phase #2 Continuous Flow (Plant Wide)				
Lean manufacturing consultant	1	40 days	$1,200	$48,000
Lean project coordinator	1	40 days	$200	$8,000
Lean project team	1	110 days	N/A	$35,000
Hourly coordinator	1	110 days	$103	$11,330
Training—hourly employees	500	15 hours	no cost	$0
Implementation hours (overtime)	500	32 hours	no cost	$0
Data analysis	20	110 hours	no cost	$0
Painting, cleaning, etc.	20	50 hours	no cost	$0
Maintenance/tooling support	10	50 hours	no cost	$0
Miscellaneous expenses (boards, cards, paint, tape, etc.)	N/A	N/A	N/A	$7,500
Outside expense—equipment relocation	N/A	N/A	N/A	$50,000

Table 5-5. (continued)

Item	Number of People	Number of Days/Hours	Rate	Total Cost
Phase #2 Continuous Flow (Plant Wide)				
Total cost—phase #2				$159,830
Savings (low)				$614,715
Savings (high)				$1,125,567
Phase #3 Standardized Work (Plant Wide)				
Lean manufacturing consultant	1	40 days	$1,200	$48,000
Lean project coordinator	1	40 days	$200	$8,000
Lean project team	1	165 days	N/A	$52,500
Hourly coordinator	1	165 days	$103	$16,995
Training—hourly employees	500	15 hours	no cost	$0
Time for implementation (overtime)	500	32 hours	no cost	$0
Data analysis	20	165 days	no cost	$0
Painting, cleaning, etc.	10	25 hours	no cost	$0
Maintenance/tooling support	5	50 hours	N/A	$0
Miscellaneous expenses (boards, cards, paint, tape, etc.)	N/A	N/A	N/A	$3,000
Outside expense—equipment relocation	N/A			$50,000
Total cost—phase #3				$178,495
Savings (low)				$737,658
Savings (high)				$1,350,681

Table 5-5. (continued)

Item	Number of of People	Number of Days/Hours	Rate	Total Cost
Phase #4 Pull System (Plant Wide)				
Lean manufacturing consultant	1	40 days	$1,200	$48,000
Lean project coordinator	1	40 days	$200	$8,000
Lean project team	1	165 days	N/A	$52,500
Hourly coordinator	1	165 days	$103	$16,995
Training—hourly employees	500	10 hours	no cost	$0
Time for implementation (overtime)	500	40 hours	no cost	$0
Data analysis	20	165 hours	no cost	$0
Painting, cleaning, etc.	5	10 hours	no cost	$0
Maintenance/tooling support	5	25 hours	no cost	$0
Miscellaneous expenses (boards, cards, paint, tape, etc.)	N/A	N/A	N/A	$3,000
Systems support	N/A	N/A	N/A	$30,000
Total cost—phase #4				$158,495
Savings (low)				$737,658
Savings (high)				$1,350,681
Total Cost of 4 Phases				**$664,650**
Savings (low)				$2,458,860
Savings (high)				$4,502,269

Table 5-6. Anticipated improvement during implementation of first four phases (aggressive schedule—2.5 years)

Category	Current Cost	Potential Improvement % (Plant Average)	Potential Savings (Low)*	Potential Savings (High)**
Finished goods inventory total $16,450,793 calculated at 15%	$2,467,619.00	-25% to -50%	$616,904.72	$1,233,809.50
Work-in-process inventory total $1,079,916 calculated at 15%	$161,987.00	-25% to -50%	$40,496.75	$80,993.50
Raw material inventory total $8,367,503 calculated at 15%	$1,255,125.00	-25% to -50%	$313,781.25	$627,562.50
Plant labor cost (plus fringe)	$33,634,560.00	-25% to -50%	$8,408,640.00	$16,817,280.00
Scrap cost (per year)	$265,000.00	-25% to -50%	$66,250.00	$132,500.00
Rework (estimate 3% labor)	$200,000.00	-25% to -50%	$50,000.00	$100,000.00
Machine/line downtime %		-10% to -50%	$0.00	$0.00
Equipment maintenance	$6,200,000.00	-10% to -20%	$620,000.00	$1,240,000.00
Material handling equipment	$110,000.00	-50% to -75%	$55,000.00	$82,500.00
Floor space		-25% to -50%	$0.00	$0.00
Past-due orders	24%	-50% to -75%		
Returned goods cost	$470,000.00	-10% to -15%	$47,000.00	$70,500.00
Employee morale (absenteeism)	1%	-5% +		
Safety (lost work days 2000)	127	-10% to -20%		
Safety (injuries 2000)	82			

Table 5-6. (continued)

Category	Current Cost	Potential Improvement % (Plant Average)	Potential Savings (Low)*	Potential Savings (High)**
Customer satisfaction	1,000 ppm	50%		$3,057,771.89
Delivery reliability	94%	25%+		
Grand totals	$44,764,291.00	N/A	$10,218,072.72	$20,385,145.50
Projected Savings Per Phase†				
Phase #1 (approximately 6 months)		15%	$1,532,710.94	$3,057,771.89
Phase #2 (approximately 6 months)		25%	$2,554,518.24	$5,096,286.48
Phase #3 (approximately 9 months)		30%	$3,065,421.89	$6,115,543.77
Phase #4 (approximately 9 months)		30%	$3,065,421.89	$6,115,543.77

† Phase #5 requires an additional 1–2 years and improves overall results by up to 20% additional
* Potential savings on the low side would be achievable with a modest effort (less than industry averages)
** Potential savings on the high side are the average for all plants implementing lean

cost, improvements in those areas can be verified. There may be improvements in the safety results, but since the cost for the current safety situation was not used to justify the project, there is no need to determine the benefit of improvement. Also, inventory was valued at 15% of the total cost, which is typical for many companies. Notice the fairly high cost of equipment maintenance—a good opportunity for reduction using lean.

Cost Analysis (Example Two)

In the example shown in Table 5-7, the company chose to absorb labor costs for implementation. These included the cost for project leaders and the project team. In addition, they chose to absorb all costs for training and implementation. Note that the return on investment is between 5:1 and 10:1. That is a very compelling reason to invest in lean.

CONCLUSION

There is always a benefit to using lean manufacturing methods. This is true regardless of where the implementer is along the journey. As long as single-piece flow, zero inventory, and zero defects are primary objectives, there are opportunities for improvement. Developing a business case is an important tool for verifying the extent of an opportunity and confirming it with other leaders within your company. After a period of time, you will begin to recognize additional improvements not apparent during the initial development of the business case. At that point, perhaps, an organization's project managers and lean implementers will more fully understand the value of operating lean. Then they will continue to drive improvements without the need to "cost justify" everything. After the initial implementation, the cost of continuous improvement is generally much lower, and the increments of improvement are more gradual. Most companies can manage gradual improvements without needing additional capital expenditures. Capital expenditures will still be needed, but they will typically be at similar levels prior to lean.

Table 5-7. Cost analysis example two—anticipated return on investment

Item	Number of People	Number of Days/Hours	Rate	Total Cost
Step #1 Current State and Future State Mapping				
Lean manufacturing consultant	1	9 days	$1,500	$13,500
Consultant travel expenses	1	9 days	$300	$2,700
Total cost step #1				$16,200
Total savings step #1				$0
Return on investment				0%
Phase #1 Stability Implementation (Plant Wide)				
Lean manufacturing consultant	1	60 days	$1,500	$90,000
Consultant travel expenses	1	60 days	$300	$18,000
Training—hourly employees	25	15 hours	$22.00	$8,250
Implementation hours (overtime)	25	32 hours	$33.00	$26,400
Data analysis	2	110 hours	$33.00	$7,260
Painting, cleaning, etc.	25	8 hours	$33.00	$6,600
Maintenance/tooling support	10	50 hours	$33.00	$16,500
Miscellaneous expenses (boards, cards, paint, tape, etc.)	N/A	N/A	N/A	$7,500
Outside expense—equipment relocation	N/A			$30,000
Total cost—phase #1				$210,516
Savings (low)				$1,532,711
Savings (high)				$3,057,772

Business Case Development

Table 5-7. (continued)

Item	Number of People	Number of Days/Hours	Rate	Total Cost
Phase #2 Continuous Flow (Plant Wide)				
Lean manufacturing consultant	1	40 days	$1,500	$60,000
Consultant travel expenses	1	40 days	$300	$12,000
Training—hourly employees	500	15 hours	$12.93	$96,975
Implementation hours (overtime)	500	32 hours	$19.40	$310,400
Data analysis	20	110 hours	$12.93	$28,446
Painting, cleaning, etc.	20	50 hours	$19.40	$19,400
Maintenance/tooling support	10	50 hours	$21.04	$10,520
Miscellaneous expenses (boards, cards, paint, tape, etc.)	N/A	N/A	N/A	$7,500
Outside expense—equipment relocation	N/A	N/A	N/A	$30,000
Total cost—phase #2				$575,241
Savings (low)				$2,554,518
Savings (high)				$5,096,286
Phase #3 Standardized Work (Plant Wide)				
Lean manufacturing consultant	1	40 days	$1,500	$60,000
Consultant travel expenses	1	40 days	$300	$12,000
Training—hourly employees	500	15 hours	$12.93	$96,975
Time for implementation (overtime)	500	32 hours	$19.40	$310,400
Data analysis	20	165 days	$12.93	$42,669

153

Table 5-7. (continued)

Item	Number of People	Number of Days/Hours	Rate	Total Cost
Phase #3 Standardized Work (Plant Wide)				
Painting, cleaning, etc.	10	25 hours	$19.40	$4,850
Maintenance/tooling support	5	50 hours	$21.04	$5,260
Miscellaneous expenses (boards, cards, paint, tape, etc.)	N/A	N/A	N/A	$3,000
Outside expense—equipment relocation	N/A	N/A	N/A	$30,000
Total cost—phase #3				$565,154
Savings (low)				$3,065,422
Savings (high)				$6,115,544
Phase #4 Pull System (Plant Wide)				
Lean manufacturing consultant	1	40 days	$1,500	$60,000
Consultant travel expenses	1	40 days	$300	$12,000
Training—hourly employees	500	10 hours	$12.93	$64,650
Time for implementation (overtime)	500	40 hours	$19.40	$388,000
Data analysis	20	165 hours	$12.93	$42,669
Painting, cleaning, etc.	5	10 hours	$19.40	$970
Maintenance/tooling support	5	25 hours	$21.04	$2,630
Miscellaneous expenses (boards, cards, paint, tape, etc.)	N/A	N/A	N/A	$3,000
Systems support	N/A	N/A	N/A	$50,000

Table 5-7. (continued)

Item	Number of People	Number of Days/Hours	Rate	Total Cost
Phase #4 Pull System (Plant Wide)				
Total cost—phase #4				$623,919
Savings (low)				$3,065,422
Savings (high)				$6,115,544
Total Cost of 4 Phases				**$1,974,830**
Savings (low)				$10,218,073
Savings (high)				$20,385,146
Total Project Cost (2.5 years)				
External resource support—350 days				$612,500
External resource travel expenses (estimated)				$70,000
Expense for carts/dollies				$60,000
Expense for miscellaneous materials (boards, cards, paint, tape, etc.)				$20,000
Equipment relocation expense				$100,000
Total				$862,500

Remember the 80/20 rule. Planning and preparation should take no more than 20% of available time. The business case normally can be completed during a two- or three-week current-state mapping process. Do not get bogged down collecting data to show every dollar. Experience shows that regardless of how detailed the data is, there are always discrepancies. The goal is to be as accurate as possible within a reasonable period. Savings will be more favorable than you predict if the implementation is at least moderately successful.

REFERENCE

Womack, James P. and Jones, Daniel T. 1996. *Lean Thinking: Banish Waste and Create Wealth in Your Corporation.* New York: Simon and Schuster.

Change Management in Lean Implementation   6

By David Stewart

Awareness of and attention to change management principles and the dynamics of change are necessary for a successful transition to a lean environment. When changes are made to an operation, project managers and lean implementers must be sensitive to change issues because most employees have not worked in a lean environment. Also, behavioral changes are required as lean plants set different expectations for performance and emphasize values different from traditional manufacturing. What worked in the old system probably will not work in the lean production system.

This chapter details an approach to change management that can be used in both brownfield (existing facility) and greenfield (new facility) implementations. It fully integrates with the five-phase model of implementing the lean production system (stability, continuous flow, standardized work, pull production, and level production), and focuses attention on the human resource and work systems. The model for change presented in this chapter is general at best and must be tailored to each situation. In particular, the history of an organization's previous change initiatives must be factored in, as their success or failure influences the receptivity to any new efforts to change the organization. Since many organizations have been subjected to different initiatives over the years, consideration should be given as to how this effort will be distinguished from all the rest.

Companies can have a poor record of learning from past experience. Despite a history of failed change initiatives, new efforts are mounted every day without regard to what went wrong in the past. Making use of lessons learned from previous initiatives (successes and failures) can ensure the success of a lean conversion. A review of some of the more common causes for failure can be helpful in anticipating problems down the road.

WHY CHANGE INITIATIVES FAIL

There are many reasons why change initiatives are not embraced and end in failure, but a few common themes inevitably lead to disaster:

- There is a lack of focus. Change is a process that unfolds over time. It involves employees learning new behaviors and giving up familiar patterns, and often is an uphill effort. Without a persistent focus on the goal, it is too easy to give up.
- There is no compelling case for change. Every worker who is asked to change wants to know why. Any change initiative must have clear and compelling reasons. Without a connection between the actions taken and the reasons given, the effort will fail to reach its potential.
- Sponsorship is lacking. When a new program begins, one of the first questions everyone asks is who is behind it. The higher the authority behind the effort, the more people are willing to be inconvenienced. If there is a perception that an organization's leaders do not support the change, then the best one can hope for is compliance without genuine buy-in.
- There is a rush to implement without sufficient development. This is an all-too-common approach to change. New initiatives are not thought through, and proper planning is forsaken to rush the implementation. However, without thought and planning, the results are likely to be short-lived.
- There is a misalignment among different stakeholders. A crucial step when planning for change is to create an alignment among all stakeholders. If this step is not addressed, fragmented efforts and conflicting goals can bring the project down.
- There is a low commitment and follow-through. Everyone involved in the change effort must have at least a minimal

commitment to its success. Adequate communications and training are critical to bringing everyone on board. If participation is half-hearted, then it cannot be expected to remain during the long haul.

FACTORS FOR A SUCCESSFUL CHANGE

Change efforts fail for many reasons, but the important factors to success are few. The following issues are the most important when managing a lean conversion.

- There must be a clear purpose. Focus, sponsorship, and a reason for change are recognized as prerequisites of the initiative. Each is an element of purpose, the point of it all. Being clear about goals and specific about desired results can help overcome many forms of resistance. The purpose must be well-communicated and hold an appeal to both head and heart.
- Effective leadership is essential. Change leadership means making sufficient resources available, providing direction and guidance, and making change a top priority. Even the process of developing the leadership team and change agents helps result in effective leadership. Making a front-end investment during the preparation stage provides leadership when it is needed most.
- Establish a comprehensive change plan and process. During the planning process, detail the change effort's scope and focus into a plan. Set boundaries and gather commitments for support. Contemplating the extent of the undertaking can be daunting, but it is important that the scope not be restricted. Lean conversions require changes throughout the organization, and the plan and effort must take this into account.
- Secure commitments to a systemic effort. Making a change from mass to lean production cannot be approached in a piecemeal fashion. It may be tempting to go after the "low-hanging fruit," selecting just those tools that seem most helpful, but lean production is a system of methods and processes that do not integrate into the traditional production approach. Since the approach is different and the changes are significant, the required commitment needs to be equivalent.

Reflect on your experiences with change initiatives in your organization. What worked and what did not? What were important factors in the successful efforts? What characterized the initiatives that ultimately failed? Do the issues mirror those previously discussed? When we ask people about their experiences with change initiatives what we hear back always confirms the importance of the preceding success factors. Given the importance of these factors in planning and design, and ultimately in implementation, a change initiative should incorporate them to ensure success.

STAGES IN THE CHANGE PROCESS

A methodical change effort goes through different stages of implementation. It is most important to spend time in planning and design before the implementation itself. Sufficient preparation is critical to a complicated and comprehensive change initiative like a lean conversion. Moving forward without preparation ensures that there will be resistance to the change, which reduces the chances of success.

Three different stages in the change cycle can be discerned and described: preparation, design, and implementation. The preparation and design stages are initiated before the implementation is launched. Within the implementation stage, there are discrete preparation and design phases that arise as different change projects begin. In the five-phase implementation process, the preparation and design stages within the change process are initiated before the stability phase begins. Once implementation of the production system is launched, the lean implementer finds that, although much of his or her time will be spent implementing, preparation and design activities are needed as the next implementation stage is anticipated. As a result, there are preparation and design phases for each new system or method implemented in the lean production system, regardless of which of the five phases the company is in.

Goals for the Preparation Stage

The first goal during the preparation stage is to establish a strong foundation for change. This responsibility lies with an organization's leaders; without their commitment to build a strong foundation, the effort will falter. Four tasks are essential at this stage:

1. Define the purpose, goals, and boundaries of the change.
2. Secure clarification and consensus on the need for change.
3. Define a shared vision of the desired future.
4. Design a change project work plan.

Define the Purpose, Goals, and Boundaries of the Change

Once the leadership decides to undergo a change effort, the purpose and goals for the initiative should be defined. Leaders' agreement on the reason for change is important, and specific goals and limits must be agreed upon. When these agreements are achieved, they can be shared with the rest of the organization. For a comprehensive, systemic effort such as a lean conversion, it is not enough to simply say the entire organization will be involved. At the beginning it is sufficient to convey the fact that a large-scale change effort will be undertaken. As the plan is developed, more specific details need to be developed and communicated about what can be expected.

Secure Clarification and Consensus on the Need for Change

The purpose of the change should relate to the business case—the reason for change. Leaders and project managers must be clear about the purpose of the change initiative, and then seek consensus from the rest of the organization. Lean conversions are often initiated due to competitive pressures that threaten a product line or the company. Making that case clear to all concerned, then explaining how lean will make things better, goes a long way toward developing the crucial alignment needed for this level of change.

Define a Shared Vision of the Desired Future

Leaders must define a shared vision of the desired future. Leadership, by definition, should be visionary. This is particularly important during a time of change. If leaders are asking everyone to make stressful and disorienting changes, the presence of a clear and detailed vision of the completed situation helps people through the transition.

It is not enough to just create a vision and announce that everyone should buy into it. Input into the specifics of the vision should be sought from those most affected by the change. Employees' opportunity to contribute to developing the vision enhances the level of commitment during the change process. The lean vision must be understood and embraced by those at the top first, then communicated to everyone else. Training and communication thus become extensions of the visioning process.

Design a Project Work Plan

The work plan for the change process also is the responsibility of organizational leaders. A road map must be designed to detail how the conversion will unfold. Using the five-phase model as a template, an organization can chart the means by which they make their conversion to lean. Overseeing and refining the work plan remains the leaders' responsibility, often undertaken by the change team or steering committee (a group of leaders that oversees and manages the change effort and coordinates the overall conversion).

Reaching Consensus

Once the goals have been set on the highest level, members of the organization must become engaged and aligned with the effort. This occurs through communicating and seeking input in the process and building alignment throughout the organization. Agreements must be made on:

- roles and responsibilities of those involved in the change effort;
- relationships and accountabilities;
- a compelling business rationale for change;
- a work plan for implementation, and
- measurables to document progress.

As the stakeholders find agreement on each of these matters, alignment is built throughout the organization for the impending implementation. Building alignment at the beginning goes a long way toward diminishing the resistance encountered during implementation.

Decision-makers must seek support from those affected by changes to move forward. However, this process of education and communication is often neglected. Those who design the change initiative should remember the adage that people do not resist change, they resist being forced to change. Attention to the change process at the level of those who must change their behavior and mindset is the secret to dealing with resistance.

Specific tasks that assist in developing support include:

- educating the organization about change plans and processes;
- gaining commitment to support the initiative from leaders above and below the leadership team;
- selecting and developing change agents;
- developing an organization-wide communication plan, and
- preparing people for change.

Once a strong foundation for change is established, alignment from engaging the stakeholders has been built, and support by educating and communicating the need for change has been developed, it is time to move to the design stage.

Goals for the Design Stage

During the design stage, the shop floor gets more active, nailing down specifics of the change process and agreeing on crucial issues. The design process is initiated by an assessment. Mapping the current state of the product through an area is required (for a detailed discussion on current and future state mapping, refer to Chapter 4). Map the product flow using conventional symbols. This will reveal areas of value and waste. This procedure alone goes a long way toward demonstrating the need for improvement.

A future state vision is then created, consistent with the organizational vision that details what the value stream might look like after improvements. Comparing the current state with the future state then reveals the needed changes.

At this point, the design process begins in earnest. Choose one area among all the others needing improvement as the initial application area, and develop a work plan for changes to that area. This area-specific work plan must be coordinated with the master plan. There should also be a plan for evaluating progress and measuring improvement.

Plans must be more than pieces of paper to be effective. Everyone involved in implementing the plans must buy into them and agree to their roles and responsibilities. If there is agreement on the need for change, then an agreement on the plan for change should follow. Begin training once there is agreement on the roles and responsibilities during the transition.

Before beginning the implementation, check the readiness for change. The questions in Table 6-1 will help determine if a company is ready to move to the implementation stage. If any question cannot be answered with a full affirmative, proceed at your own risk.

Goals for the Implementation Stage

The implementation stage encompasses more than just putting changes in place. Instituting changes on the shop floor initially puts more stress and challenges on workers as they transition to new ways of doing things and altered expectations. At the same time, there is often an effort to maintain organizational performance (for example, production will not suffer). This may be an impossible expectation; some understanding needs to be brought to bear. Close tracking and coordination of the implementation is important, since people often feel more out of control during a transitional period. The impact of the changes on production,

Table 6-1. Checklist for change readiness

Is leadership/sponsorship in place?
Are the goals clear?
Are the roles and responsibilities clear?
Are sufficient resources allocated?
Is there a compelling business rationale that all can understand?
Is there a clear and compelling vision of the future state?
Are stakeholders committed to the change?
Are the recognition and reward systems congruent with the change?
Is there a way of documenting progress (measurables)?

morale, and safety should be closely monitored. Pay attention to the psychological dynamics of change to help deal with much of the resistance that occurs during this period.

Realize that giving up the "old" ways of doing things represents a loss of security and assurance. This can lead the lean implementer to anticipate and be in a good position to help employees respond to their loss. New behaviors are not yet in place, so there is anxiety about being able to excel at new methodologies and processes, as well as the old ways. Providing support by rewarding progress and learning becomes all the more important to speeding through the transitional period. Continued training, leadership support, and communication are critical to workers implementing changes during this period.

Evaluate results of the changes after they have been fully implemented and the transitional period has been completed. This process leads to amending implementation plans in other areas. Lessons learned should be applied to the overall change initiative. In a lean conversion, the changes follow closely on each other. The five-phase model calls for a staged process of changes. Employees rarely have a chance to achieve a sense of stability before a new change is introduced. On the plus side, benefits of the improvements help offset the stress and loss of motivation brought on by constant change. Careful monitoring is recommended.

Employees need to be recognized and rewarded for their achievements during and after a protracted change effort. To support and sustain the changes made, ongoing renewal efforts, which continue to reward and reinforce the changes in practices and behaviors, should be a matter of course after the change team moves to another application area.

ROLES AND RESPONSIBILITIES

During a change initiative, some roles are critical. They must be clarified and agreed on to move forward. Roles, and the responsibilities that go with them, are created and supported at the highest level of the organization. As emphasized earlier, organizational leadership plays a critical role in the conversion to lean. In the process of sponsoring and leading change, senior managers have specific responsibilities that cannot be delegated elsewhere. Supporting and sponsoring the effort in the beginning and throughout the conversion process are the most important duties. In the beginning, leadership defines the need for change and creates the vision of how things can be different. As others in the organization begin to embrace the vision, leaders must continue to communicate the need for change and support the efforts being made to realize the vision.

The steering committee is intimately involved at the strategic level, creating the plan for change and dealing with specific issues that arise during design and implementation. This group often has representatives from areas affected by changes. In union plants, having union representation is important to signal a cooperative spirit. Members of the steering committee will live and breathe the changes on a daily basis throughout the process, so it is best if they have some relief from their normal responsibilities.

Individuals on the shop floor who are most actively involved in the implementation process are the change agents and implementation teams. Teams may vary in composition from one area to the next. Make sure they are committed to lean conversion and can promote the initiatives, as they will deal with resistance to change on a daily basis. Training is necessary before they start to work on the shop floor. A summary of the roles and responsibilities is shown in Figure 6-1.

RESISTANCE TO CHANGE

The most significant obstacles when making the transition to lean involve the expectations, values, attitudes, and priorities that employees bring with them. Collectively summarized as "mindset," the effect these factors have on promoting rather than hindering change is frequently underestimated. Existing mindsets are unconsciously maintained, so they are not easily changed by exposure to new ways of doing things. Experience in traditional shop-floor environments creates skepticism about lean tools and processes. This manifests itself as a reluctance to try, much less embrace, new methodologies, even though there may be a conscious acknowledgement of their superiority. In this case, to listen

Change Management in Lean Implementation

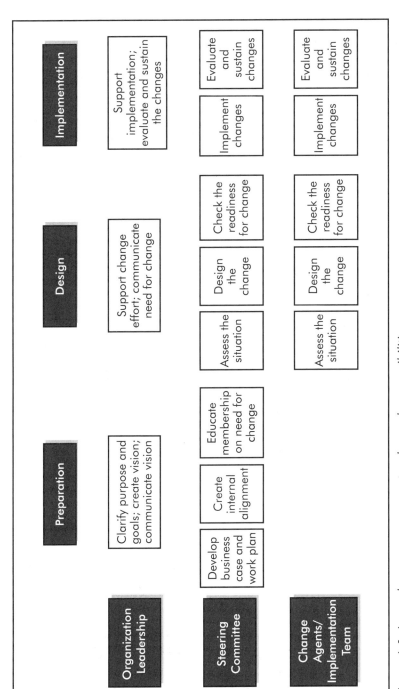

Figure 6-1. Lean change management: roles and responsibilities.

for objections and address them in a respectful manner is most productive. Old shop hands can do lean if their reluctance and objections are dispelled in a courteous, thoughtful way.

The following items are some of the challenges and outcomes represented by changes in mindset.

- There is a transition from command-and-control leadership to participative management. The most common style of management in traditional manufacturing environments is a supervisor exerting and expecting maximum control over his or her "troops." This "command-and-control" style is effective when input or involvement from the operators is not needed and morale is not a concern. A directive style from a leader who knows what needs to be done is appropriate in times of crisis. However, lean plants typically do not function in crisis mode, so input and involvement from the line is an important part of the culture and system. While directive management is called for at times, participative styles are best for encouraging involvement from subordinates. Learning how to coach, facilitate, support, and encourage becomes a critical part of re-training a traditional manager. Attention to this important mindset issue is a critical variable in the lean change effort.
- There is a change from rewarding individual achievement to rewarding teamwork. Imagine trying to encourage people to work together in a harmonious and collaborative fashion while continuing to reward individual achievement and heroic efforts. Confusion would be the most likely result. Lean plants require that employees work in groups or teams in which the common good is more important than individual heroics. Managers unfamiliar with team-based environments must learn and discover ways to support and encourage the cooperative spirit, without discouraging individual effort.
- The work mindset shifts from maintaining the status quo to continuous improvement. The status quo is the comfort zone in most traditional manufacturing environments. Doing things as they have always been done is more important than innovation or change. The lean system encourages the reverse. One is never satisfied with the latest innovation; instead, the standard is used as a baseline for improvement.

Cultivating an appreciation for positive change can be a significant mindset challenge. It calls for a different approach to problems than what is customary.
- There is a change from hiding mistakes to identifying problems and active problem-solving. For employees to acknowledge mistakes and raise problems, they must be able to trust that they will not be punished for stating the concern. Line workers in a lean plant must know that if they pull the Andon cord to stop the line, they will not be punished. Active problem-solving starts by rewarding the discovery of problems. Managers not familiar with seeing problems this way must learn new behaviors that often feel contrary to what they have always known and believed. This mindset challenge can be overcome with training and coaching.
- The competitive environment becomes a collaborative workplace. Just because the marketplace is competitive does not mean the internal organization needs to be competitive. The lean enterprise is built on communication, cooperation, and responsiveness. Rather than viewing the supplier or customer as a problem, seek ways to improve collaboration. This is especially true within the plant, between internal suppliers and customers. Smooth coordination between areas and functions supports the continuous flow that is central to the lean approach. Traditional managers find this emphasis to be at odds with their previous experience. They must find ways of adapting to the collaborative workplace.
- There is a transition from specialization to the multi-skilled worker. Many behaviors required by the change effort need to be sustained in the new lean system. Communicating, aligning, and coordinating efforts is essential in the transition as well as when the production system is established. In this sense, the change effort is the first test of an organization's ability to sustain the behaviors needed for lean. Incomplete or failed transitions likely reflect a lack of readiness or ambivalence about the nature of the changes required. Problems in organizational change often reflect the difficulties that individuals have in embracing the changes.

Leaders must be fully behind the changes in a public and uncompromising way. The vision for lean must be directed from the top and communicated throughout the organization. Lean production systems require more from the front-line worker than traditional mass production, so more decision-making and control must be relinquished by managers. Also, a more participative style has to replace the conventional command-and-control approach. Leadership styles that foster teamwork have to be cultivated—coaching, supporting, and delegating should supplement directing as management alternatives when working with shop floor workers. A more disciplined, fact-based, problem-solving methodology replaces the usual fire-fighting that is the supervisor's daily fare.

- There is a change from traditional management philosophy to lean principles. The proper lean mindset required of management includes changes from functionally autonomous to interdependent; from lone experts to a disciplined process; from the use of experience and instinct to the use of data, and from separate functions to a systems mentality.

Some lean themes reflect necessary changes in thinking and approach. The customary silo mentality ("it's not in my area"), must give way to a systems approach ("how can we eliminate waste?"). Everyone assumes responsibility for customer satisfaction in a lean organization. Just as individual responsibility is encouraged, collaboration and teamwork is valued over individual heroism. Changing to a work group, team-based organization brings challenges in and of itself. (See Chapter 10 for a full discussion of work groups.)

CONCLUSION

A lean initiative can often seem intimidating when considering the range and magnitude of changes to shop floor practices, management behaviors, and organizational structure. To counter the sense of being overwhelmed, it is important to have a specific change management approach that provides a way of bringing structure to all the changes required. The approach described in this chapter is based on experience with lean implementations in a variety of settings and should serve as a general template for thinking through the various considerations that arise.

This chapter, along with those on the five-phase implementation process (Chapter 7) and creating the human system organization (Chapter 8) should provide the implementer with the necessary maps to begin planning specific approaches to the lean destination.

The Five-phase Implementation Process 7

By *John Allen*

When company leaders decide to implement lean manufacturing, a template of the changes needed and outcomes desired is useful in organizing the steps in the process. A highly successful template for the implementation of lean manufacturing is the five-phase model, which was first introduced by Toyota. The five-phase model breaks down the critical elements of lean manufacturing and places them in a sequence to be followed during an implementation process. Knowing what comes first and what follows is extremely helpful during the planning of the implementation, as well as during the implementation itself. This knowledge is especially helpful when questions arise about the desired outcomes. The five-phase model provides the blueprint for building the house of lean.

This chapter introduces the five-phase implementation model, providing an overview of the pertinent elements at each stage and issues that commonly arise during the implementation effort. The chapter is organized around the major components of a lean company: quality, machine reliability, material utilization, and people systems. Each of these components must be firmly in place for the production systems to be truly lean.

Each organization is different, so the strategy for becoming lean must be specific to each company's individual needs. The "one-size-fits-all" mentality is not useful in creating a lean organization. Whichever approach to lean implementation is chosen, it must be flexible enough to accommodate the organization. For example, if you have had a major, successful effort in your organization to enhance equipment reliability, it is not necessary to re-establish that foundation.

PREPARATION

A successful lean conversion requires detailed completion of all preliminary steps, such as mapping the current and future state, developing a business case, and organizing change management activities. Although change management is discussed in detail in Chapter 6, it is worth exploring the difficult, yet important task of creating a vision that achieves commitment, alignment, and action toward the desired future state. Creating a vision is difficult because implementers believe they need a "perfect" description of what the future looks like. It is unrealistic to expect that a non-lean organization can create a descriptive vision of the future. It is the process that creates the all-important synergy of effort, not the preciseness of the vision statement. A vision statement is a work in progress.

To alleviate fears that the organization may be chasing unrealistic dreams, create the initial vision for no more than one year into the future. As time goes by and the clarity of purpose becomes apparent, the vision's duration can be lengthened to seven or 10 years. This way, the vision becomes something real and tangible for the organization.

Changing the vision is desirable and necessary as the organization learns from the implementation process. One shortcut in the discovery process is to engage someone with experience in implementing lean to detail the necessary conditions. Even with this expert, however, an organization must be careful not to let the visioning effort be the product of expert knowledge. Those in the organization who will implement the lean concepts, the key stakeholders, should produce the process and resulting outcomes.

When all the preparatory activities are completed, the organization is prepared to begin implementation. Implementation must be well-organized and precisely executed so the right signal is sent to the rest of the organization. Shop floor personnel, managers, and others are most likely waiting for affirmative leadership. The leaders, however, have not had any experience implementing lean. Providing solid leadership can be difficult unless implementation architecture exists to instruct those involved. Architecture must be simple, easily understood, and provide a picture of how the implementation will proceed.

THE FIVE-PHASE MODEL

The architecture that best fits the preceding requirements is called the *five-phase implementation model*. It was first introduced to American industry by the Toyota Supplier Support Center as it struggled to create a process to teach the Toyota Production System to suppliers. The following phases have been modified by the authors to more accurately reflect typical American industry norms. It has been flexible enough to handle every implementation scenario encountered to date. The model has been used in more than 50 different businesses in at least 15 industries, with organizations from 150 to more than 300,000 employees. In every case, the five-phase implementation model has brought positive results. The degree of success, however, depends greatly on individual circumstances and the leadership directing the implementation.

The Model Defined

To begin to understand the five-phase model, it is important to detail the expected outcomes. The common denominator when detailing outcomes is waste elimination. Waste is the difference between an organization's current state and the future state of being lean, which incorporates an active process of continuous improvement. Specifically, there are four outcomes:

1. Material and processes deliver to the customer exactly what is needed, when it is needed, with the quality expected (the "Just-in-Time" philosophy).
2. The lean system is able to support a smooth flow of product through operations. Product flows without stopping for any reason, and value is added at each step (built-in quality).
3. Employees are able, flexible, and highly motivated to satisfy the customer.
4. Equipment is reliable and available, with processes to provide constant feedback to the machine builder to increase reliability in the future (operational stability).

Together, these outcomes create the major elements of a "house" of lean manufacturing, in which all serve in delivering value to the customer (see Figure 7-1).

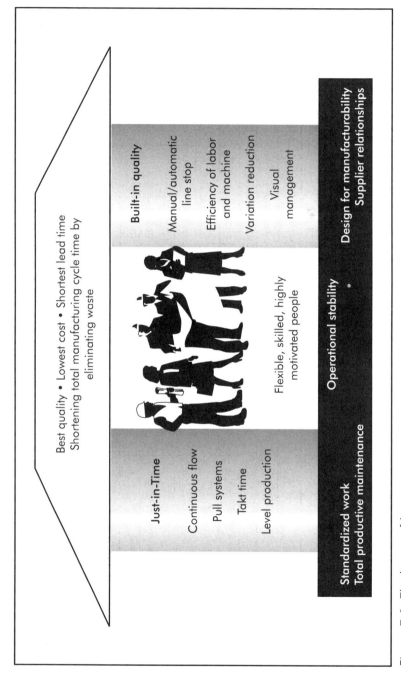

Figure 7-1. The house of lean.

In Figure 7-1, the two main pillars of built-in quality and Just-in-Time rest on a foundation of operational stability held together by flexible, skilled, and highly motivated people. Built-in quality control is emphasized by ensuring that nonconforming parts do not leave the workstation. Just-in-Time ensures that the right product (at the right cost and in the right quantity) arrives to the customer. Reliable processes and equipment can deliver the highest quality. People can create change or resist it, but learning to deal with them constructively enhances the chances for success. These issues must be addressed in the quest to achieve lean manufacturing through defined steps. The steps are characterized as the five-phase implementation process. Initial phases serve as a foundation for the following phases. Completing all of the steps correctly achieves the results necessary to implementing the overall system. The five phases are defined as:

1. stability,
2. continuous flow,
3. standardized work,
4. pull production, and
5. level production.

The Five Phases

The following outline defines each phase and describes the objective at that stage of the implementation. Changes that can be expected after the objective is achieved are detailed, as well as the lean tools used to help achieve the goal. To better understand the resistance encountered at each phase, detracting and enabling forces are described, along with the leadership required to overcome these obstacles.

Stability Phase

Definition. *Stability* is knowing when instability exists and how to react to eliminate it.

Objective. Implementers must find sources of instability and create the capacity within the organization to diagnose and eliminate them.

Expected changes. The transition is made from a traditional organization to one in which everything has a place and everything is in its place. In addition, a concerted effort is made to improve equipment uptime, and understand the tasks necessary to make a quality product. Relationships between supervisors and operators are defined and improved.

Tools to be used. Required tools include 5S; visual factory; constraint analysis; equipment repair; standardized work (minus Takt time analysis); point-of-use tools; statistical process control; organizational commitment building; lean training; current and future state mapping, and work group initiation.

Enabling forces. Forces on the plus side include management commitment; cleaning equipment for inspection; implementation teams; steering committee; a quality operating system, and human resource systems.

Detracting forces. These forces include failure to focus on the people who add value; lack of management support; implementation without a plan, and failure to create sufficient understanding for those who are implementing.

Leadership required. Leadership is required to understand the short-term vision; agree on measurables for the shop floor; understand where to begin; supply the organization with the style of leadership required, and put together a successful implementation team.

Continuous Flow

Definition. *Continuous flow* creates processes that flow smoothly through all the operations without stopping.

Objective. The objective in the continuous flow phase is to define why processes are not flowing continuously and alleviate the bottlenecks.

Expected changes. Implementers can expect a change from a process that stops for repairs, lack of parts, improper methods, and changeovers, to a process that flows smoothly from workstation to workstation.

Tools to be used. Necessary tools include mistake proofing; quick changeover; conveyance minimization; TPM; flow racking; leveling production on the basis of volume, and an aligned manufacturing organization.

Enabling forces. Positive areas include identifying and solving problems; material planning; visual management; team involvement; policy deployment; and communication mechanisms.

Detracting forces. Forces that negatively affect the process include a lack of involvement by support organizations; lack of clear business goals; lack of involvement by the person adding value; leadership without vision; poor relationships, and inventory.

Leadership required. Leaders must provide clear goals and responsibilities; frequent monitoring; motivational communications emphasizing success, and a willingness to plan.

Standardized Work

Definition. *Standardized work* means synchronizing the product's rate of flow with the demand of the customer.

Objectives. The goal of this phase is to ensure that the rate of production is no more or no less than the rate required by the customer.

Expected changes. Implementers can expect a change from an organization that is focused on achieving flow to one that is focused on flow at the rate of customer demand.

Tools to be used. Required tools include standardized work; Takt time; work group teamwork; task re-balancing; minimum and maximum levels on flow racks; mistake proofing, and quick changeover.

Enabling forces. Forces that help implementers achieve the goal include effective work groups; disciplined task accomplishment; leveled volume; policy deployment; problem solving, and team leaders doing the initial standardized work.

Detracting forces. Forces that take away from the effort include random task organization; team leaders not monitoring task accomplishment; failing to monitor performance based on the policy deployment, and failing to elevate problems.

Leadership required. Implementation leaders are required to develop a participative leadership style; insist on following the standard; establish a span of support for the team leader, and insist on achieving the vision.

Pull Production

Definition. *Pull production* is defined as having the right parts at the right place, at the right time, in the right quantity, and nothing more.

Objective. The goal of pull production is to have the downstream operation take (pull) the items needed from the previous operation into its workstation at the moment the last product has been completed. In addition, the goal is to make materials and supplies arrive at the workstation at the moment they are required to complete the production process.

Expected changes. Implementers can expect a change from in-process storage of many parts to having one part ready for the production process (having just completed it at the preceding process).

Tools to be used. Necessary tools include fixed schedules; Kanban; daily schedule attainment; synchronous internal and external logistics; monitoring inventory levels at the workstation by the operator; disciplined production processes; and continuous improvement from the teams.

Enabling forces. Forces that promote this phase include production control by the material planning and logistics department; milk runs (a spirit of improvement originated at the point where value is added); disciplined standardized work; central storage areas, and visual control of the workstation.

Detracting forces. Forces that take away from the effort include materials resource planning; random task accomplishment; just-in-case inventory, except for safety stock; failure to respond to Andon signals, and unreliable suppliers.

Leadership required. Implementers must insist on a disciplined replenishment process; establish goals for improvement; facilitate supplier development; emphasize productivity and quality; monitor the process on a regular basis; and develop individuals as a business issue.

Level Production

Definition. Level the product on the basis of volume, mix, and sequence.

Objective. The goal of level production is setting daily production schedules; making every product every day; achieving the daily schedule each day, and getting support for obtaining consistent daily schedules.

Expected changes. Implementers should expect a change from maintaining a finished goods inventory for the customer to pulling product based on their needs. Also, they should expect running a sequence that is leveled (in volume, mix, and sequence) over a fixed period, such as one month, and equal to the customer demand.

Tools to be used. Necessary tools include daily schedule attainment; sequencing production at the beginning of assembly; running to the sequence; producing every product each day; leveling the sequence, and constant monitoring of the production process to ensure daily output.

Enabling forces. Forces promoting this phase include operations control by material planning and logistics departments; effective forecasting; synchronous subprocesses, and predictive maintenance.

Detracting forces. Issues that can take away from this effort are a lack of discipline; supporting processes that are not aligned; not meeting supplier schedules, and random replenishment.

Leadership required. Lean leaders must support productivity-enhancement activities; clearly articulate the vision so the goals are visible; insist on the measures being achieved; develop subordinates, and attend to the shop-floor environment.

Practical Considerations

During implementation, many questions are apt to surface about the decisions made, the processes used, and the outcomes desired. Although it is impossible to address every possible question that might arise, it makes sense to discuss some common issues that are likely to emerge.

Systemic Implementation

Unlike a pick-and-choose situation, lean implementation must be system-wide. The five-phase model is a rational method for what needs to happen, and is built on the assumption that implementing all of the phases affects all aspects of the system. When selecting specific tools to implement without taking a systematic approach, the lean implementer is likely to miss parts of the system. For example, if work groups are ignored at the outset, the most likely outcome is a system that cannot be properly maintained because those who are most affected lack proper commitment.

The Production System is the Culture

The classic definition of culture is that the production system sets expectations for those involved, at any time, and under any circumstance. Therefore, the production system is the culture. Failing to understand this causes organizations to chase a new culture that is either ill-timed or inappropriate (for example, the current fad to install self-directed work groups before the stability of the organization is achieved).

The Role of Management

Company managers are impatient, and they do everything in their power to ensure that their organization is operating at its highest level. As a result, they push too hard and too soon to demand tangible results. The result is that they force their organization to shortcut the process of maintaining and sustaining improvements before additional ones come online. Managers feel that a "home run" is needed, rather than building upon small successes.

Managers are responsible for providing leadership in the implementation effort. A lack of leadership can cause subordinates to place less priority on lean implementation. Ultimately, leadership does not mean having all the answers, but a willingness to learn and teach. Leading by example is the strongest form of leadership.

Leadership Alignment

As the production system sets expectations for all members of the organization, so must all levels of leadership be united to champion and sponsor the lean vision. Unfortunately, supervisors and middle managers are often the last to be engaged, leaving them with no other option but to revert to their old behaviors and subvert the overall effort.

The Importance of Details

Because of the desire to hit a home run, every time an implementer is at the plate he may fail to appreciate that the real power of implementation lies in the details. This lack of appreciation for the finer points causes implementers, at worst, to place too little emphasis on implementing standardized work. Standardized work is the key to successful implementation.

Problems are Good

Elevating problems to the proper level of management is a key factor in making progress. In many companies, employees do not relish bringing bad news to the supervisor because it makes them look weak and indecisive. Viewing problems as opportunities is non-existent. The only way to fail is to have problems and not elevate them so they can be solved.

SUCCESSFUL IMPLEMENTATIONS

In the long and storied history of lean manufacturing, only a handful of companies have achieved full implementation. In each case, the circumstances surrounding the implementation were:

- a clear vision of the company's goals;
- insistent leadership that supplanted the lack of knowledge with proper preparation, and
- monitoring the plan of implementation.

The most dramatic cases of lean implementation in which the authors were involved were New United Motors Manufacturing in Fremont, Cal., and Toyota's manufacturing plant and Johnson Controls' plant, both in Georgetown, Ky. It is a mistake to think that what worked there is the exact formula for every implementation. Careful observation of successful lean companies shows that following a great plan and insisting on measurable results are requirements. The five-phase model accounts for these factors in the implementation process. This process combines the logic of sequential steps with the expectation of measurable results. Combine these with inspiring leadership, and the implementation has its best opportunity to reach a complete state of lean.

IMPLEMENTING LEAN ELEMENTS

The house of lean in Figure 7-1 shows the major components of a lean company and how they relate to one another. Any implementation plan must take into consideration each major element of:

- developing a quality system;
- creating reliable and maintainable equipment;
- delivering materials to the place they are needed, when they are needed, and
- maturation of the people system.

These elements must be addressed during any implementation process. Short-changing any (or any combination) of these causes the system to fail.

Develop a Quality System

A quality standard provides a basis for operations in a lean organization. The absence of quality (defined by the customer) creates the need for rework, repair, and scrap—all forms of waste, but correctable. The discipline underlying quality is understanding how standards translate to the process of adding value. Once this is understood, many options are open to help arrive at zero defects. Mistake-proofing, for example, is open to the operator because he or she can distinguish between good and bad quality.

Quality is established when a product's delivery process meets customer expectations. Quality is then incorporated into the manufacturing process using standardized work principles to maintain consistency. From this point, feedback provided by the customer (internal or external) is essential to maintaining quality. Mechanisms can be set in place to test, observe, or operate the product to maintain proper quality levels. From these operations, feedback is sent to the operator indicating whether standards are lacking or quality is attained. Feedback allows the operator making the product to change the practice or reinforce good quality.

In lean terms, inspection is a form of waste, but it may be necessary to ensure that customer expectations are being met. The manager's challenge is to minimize inspection while maintaining the highest quality. An effective quality system integrates quality creation, quality maintenance, and improvement efforts. The system involves each member of the organization working in concert to fully satisfy the customer.

Stability

During the stability phase, the effort expended must establish quality levels and put into place control over design, incoming material, production, and new process development. Each of these efforts is supported by a quality engineering organization that monitors and measures performance, and provides feedback to improve the process. Company managers must determine the desired level of quality for design, product, process, and procedures. Setting high standards for quality mandates that everyone from the highest officer to production personnel work to achieve and maintain the high standards.

Making quality standards visible to everyone, especially the operator, is a major activity for lean project managers. Visibility can be as rudimentary as banners with slogans, or as involved as defining how the workplace must be organized to achieve quality levels consistent with customer expectations. The operator is the source of ideas for visual efforts. He or she must understand what constitutes quality, and have simple methods for ensuring it.

To organize the effort of setting high standards for quality, project managers must establish goals, objectives, and a target. They must practice the philosophy of quality and organize the company to reach customer-driven levels of quality. Once the goals, objectives, and targets are established, managers must make sure employees understand and are committed to achieving the standards. This task is achievable with the proper resources.

Managers also must firmly answer questions from those in the organization who will challenge whether quality is really important or just lip-service. These questions may include:

- "This product is not up to our quality standards, should I ship it?"
- "What is more important, quality or shipping?"

At the line worker level, absolute quality must be incorporated into standardized work or process documentation. Critical characteristics of the product, such as safety, critical torque, or appearance, must be visible in the initial work instruction. Later in the five-phase process (in the standardized work phase), quality standards are documented in work instructions and checked periodically (twice per shift) to ensure their compliance.

For the supervisor, maintaining quality is a primary objective. Primary objectives can include:

- adjustment checks before the shift;
- monitoring quality levels produced during the shift;
- routine critical characteristic checks;
- involving the quality department on incoming material quality levels;
- participating in writing and interpreting quality specifications for the customer, and
- studying process quality capability for new products and processes.

During the stability phase, process controls are developed to create a procedure that delivers consistent quality. These controls include identifying the critical characteristics of the product and developing processes, flow charts, characteristics matrixes, and control plans (with failure modes identified). Operators can use controls in daily operations and, ultimately, they can be included in their standardized work.

As in all manufacturing operations, lean does not preclude the process of changing product specifications from taking place. Therefore, a "manage-the-change" process must be created to ensure the most up-to-date specifications are available to the operator.

Continuous Flow

Quality, as it relates to the continuous flow phase, emphasizes correcting issues that prohibit flow. When an operation stops, the organization must be capable of determining the nature of the problem, putting in temporary countermeasures, and ultimately solving the root cause of the problem.

One methodology used to ensure that problems do not recur is to mistake-proof the process. Mistake-proofing is a tool of lean implementation. The operator, with assistance from engineering or other support organizations, designs methods to keep problems from recurring. Allocating time to work on mistake-proofing devices is a sure sign of a high level of commitment to maintaining quality.

Standardized Work

Maintaining established quality levels and integrating Takt time into standardized work are major quality issues in the standardized work phase. Both issues are foundations to creating and maintaining a lean organization. If either issue is not viewed as an absolute standard, the organization will never achieve lean.

Maintaining established quality levels requires discipline and diligence from every segment of the organization. Operators must always repeat the same sequence proven to deliver quality in every cycle. In addition, they have to understand quality in their finished product and in the product as it arrives at their workstation.

Team leaders must teach standardized work to all new or rotating workers so they can maintain quality standards. They also must diagnose quality problems and resolve them within the Takt time to maintain flow.

Supervisors must observe the process and determine if quality products are being made consistently. In addition, they must analyze data from various sources to determine whether quality is being achieved.

Managers must ensure that line performance is consistent with established quality goals and targets. In addition, they are responsible for encouraging the organization by celebrating milestones and achievement during the implementation.

Pull Production

In the pull production phase, the emphasis is to ensure that incoming quality is consistent with putting the product on the line in small quantities, which lowers inventory levels. Suppliers must promote perfect quality achievement through the production process. Failing to achieve perfect quality forces an organization to compensate by having buffer stock at the line or in a marketplace. The manufacturing organization requires immediate feedback from the production process to contain quality problems and resolve issues at their source.

It is imperative that operators make a visual determination on whether in-process stock has the necessary quality. Samples of the incoming product are necessary to make sure it meets quality standards. Defects can slip through if the level of sampling is not consistent with the level of incoming quality.

The major concern for team leaders observing the process and maintaining quality becomes finding problems that could disrupt the production process. These issues are usually found by responding to signals from an operator that a problem exists. If the issue can be resolved within Takt time, then the team leader is monitoring trends by tracking data after instituting a fix. If the issue cannot be resolved within Takt time, team leaders must seek help from the supervisor, who determines whether the quality organization must be involved.

Level Production

The level production phase emphasizes absolute insistence on in-station quality, so production can proceed in sequence with the proper volume and mix. The leveling process causes the remaining quality issues to quickly rise to the surface. The internal organization must be prepared to solve quality issues directly. If this is not done, the organization protects itself by having adequate buffer stock available to compensate for the lack of high quality standards.

Create Reliable and Maintainable Equipment

Productivity and quality are compromised if machines are not able to maintain specifications for the product. The process for ensuring appropriate specifications is complex and requires substantial involvement. It begins before the machine is purchased and continues indefinitely as the machine reaches higher levels of reliability.

For most organizations, substantially higher levels of productivity are achieved simply by improving the availability of equipment. Achieving higher levels of availability is tied directly to the phases of implementation.

Stability

In the stability phase, lean implementers must establish a preventive maintenance program for equipment based on usage, condition, or time-based criteria. Procedures must be developed, including systems for equipment feedback, lubrication, gaging and testing equipment calibration, and collecting historical data.

Once preventive maintenance processes and procedures are in place, the entire organization must support them. Create a maintenance team to work with equipment that restricts higher levels of production. Begin the work with a cleaning aimed at identifying abnormalities. While cleaning, determine lubrication points and target any source of contamination for elimination. Afterward, correct contamination sources and major losses (equipment waste) through work orders.

From the data gathered during the cleaning, develop lubrication and bolt-tightening procedures. Also, perform planned maintenance and replacement activities according to the data gathered. Maintenance and replacement activities are standard processes for ensuring compliance with equipment wellness. In addition, they serve as a baseline for equipment performance (30 days of data minimum). Gathered data is useful in setting objectives and expectations for the upcoming review period.

Continuous Flow

In the continuous flow phase, machine reliability activities concentrate on the performance of planned maintenance activities. Inspect maintenance performance frequently to establish the necessary discipline for continued compliance. Afterward, the frequency of inspections can be altered to satisfy the commitment to planned maintenance.

A program for continuous equipment monitoring is now possible. This enables greater gains in reliability while working with longer-term issues. It also provides data when working with machine vendors to create reliability through the design process. Critical processes may require a fixed schedule of continuous monitoring. Analyzing data allows project managers to create a planned maintenance review process. The improvement cycle has begun, and the continuous improvement of equipment becomes a reality.

Machine-specific, on-the-job training modules can be created and used to educate beginning operators and rcinforce lessons learned on the job. Having an established format is helpful when devising brief, single-point lessons. Making them available in a notebook or computer file on the shop floor enables frequent review by operators and maintenance personnel.

Standardized Work, Pull Production, and Level Production

Issues are consistent in the standardized work, pull production, and level production phases. The first priority is to continue to improve the gains made through the stability and continuous flow phases. The second priority is to begin refining reliability and maintainability through data analysis and refinement with machine suppliers.

Maintaining Gains

Maintaining the gains achieved through the first three phases is perhaps the most difficult issue in equipment reliability. It requires aligning the goals with the action taken to achieve the objectives. Alignment is followed with constant monitoring to ensure that the standards are not being lost during implementation.

Align goals with a process called *policy deployment.* Policy deployment is used to maintain equipment reliability, and is universally applied to all aspects of lean implementation. It is the process of aligning actions to goals and gaining commitment to secure great performance. The process begins by assessing the current state of the organization. This is done by gathering performance data on equipment uptime, performance efficiency achieved using the equipment, and product quality from each machine. Assessing the current state may be a daunting task if the data has not been collected in the stability and continuous flow phases. Compare a machine's health to the equipment requirement established under business goals and targets for improvement.

Lean implementers must establish targets and goals, and communicate them effectively so subordinates can turn them into action plans for the upcoming year. Project leaders should determine the level of subordinates' understanding by the quality of their plan. The leader then assesses the help needed to put the plan together, and ultimately determines monitoring frequency based upon his or her confidence in achieving the plan.

The amount of follow-up and monitoring done by the leader makes subordinates aware of the emphasis placed on the goals. Monitoring occurs inversely with the level of confidence that the goals are being accomplished. Monitoring can range from reviewing performance to analyzing trends and determining how they affect goal accomplishment. In every case, monitoring must be accompanied by physically seeing the situation on the plant floor.

In achieving a significant research and maintenance (R&M) program, the goal becomes attaining greater levels of machine reliability through data generated on mean-time-to-repair (MTR) and mean-time-to-fail (MTF). These are the expected times of operation without need of repair (MTR) and the average lifetime before expected failure (MTF). MTR and MTF data is used by machine designers, machine builders, and production personnel to build higher levels of MTF and lower levels of MTR into the machine.

Deliver Materials Where and When Needed

The goal of lean material utilization is to have the right amount of material arrive at the right place, at the right time, and in the right quantity (Just-in-Time). Anything more is waste and is subject to being eliminated or minimized to achieve lean goals. In the process of achieving Just-in-Time materials, all phases of the implementation are affected.

Stability

The major issue in material utilization during the stability phase is creating the capacity to handle such a large undertaking. Form a logistics planning team to collect the current state information, determine the scope of a plan, and develop a strategy and overall material flow plan. As in all cases of lean implementation, the plan needs to be reviewed by key stakeholders in the process with the aim of achieving a consensus. If this process leads only to information, but not agreement, then the plan will die from a lack of commitment. However, if the goal of the review process is understanding, this will ultimately lead to commitment by the key stakeholders.

In material flow, a key issue is space. Space requirements are driven by a trade-off from buffers taking up space to having marketplaces in consolidated areas. The space requirement in a marketplace is less, but material is concentrated in one area. This is a difficult transition on the shop floor because operators see the buffers as their only source of protection from an ever-changing requirement. Shop floor operators must see lean material flow as a process and part of a larger system, which aims to help them achieve stability.

Once a consensus among stakeholders is achieved, it becomes possible to further develop a plan for material planning and acquiring support equipment. This system monitors usage and turns it into replenishment requirements.

Continuous Flow

The largest issue to face in the continuous flow phase is reducing in-process inventory by implementing level schedules for volume and mix. This leads to the discipline of daily schedule attainment. Line operators can get behind the effort and not feel out of control. The leveling effort in continuous flow is mostly aimed at getting production to move smoothly from process to process without stopping. Material is an enabler of the process flow, but it must be carefully managed so it does not break the flow. In the standardized work, pull system, and level production phases, material reduction can be more aggressively pursued and will not undermine the implementation.

Standardized Work

In the standardized work phase, the organization synchronizes the flow of production to the customer demand. Each operation targets completion within Takt time. With this addition to the manufacturing flow, the organization can begin to implement the pre-conditions for pull manufacturing. For operations that cannot achieve continuous flow, there are market areas or stores. Market areas and stores provide enough material for the subsequent process to maintain flow at the customer's demand rate. The continuous improvement effort for stores is to reduce their size by finding ways to make the upstream process more efficient.

Pull System

The most significant effort in the pull system phase is getting the flow of material to the line to match the flow of the process. Pull is based on the customer's demand rate. It refers to both the pull of materials to the line and the pull of production from one process to another. In its perfect state, material flows to the line and is used at the exact rate in which products flow from process to process. Pull is facilitated with a Kanban card (generated by the supplier) that is entered into the system by the operator and analyzed by the production control department. Each operator signals to the others the rate at which material is to progress. Material usage and the production rate can be used to determine how much material is needed to handle the pull from process to process.

Level Production

Production flow is leveled on the basis of volume, mix, and sequence in the level production phase. Production planning can reach its most efficient point when minimizing the amount of material delivered to the line. The financial impact of reaching this level is substantial, and a foundation for even further savings is created.

Maturation of the People System

Lean implementation will never succeed unless the people system is engaged. It is imperative to make sure the people system is involved and mobilized during the implementation. Employees can either accelerate or halt the implementation effort; they can sustain the implementation merely with their enthusiasm. Paying attention to people systems is absolutely necessary for lean to become reality. See Chapter 8 for a full discussion of engaging people systems in lean implementations.

CONCLUSION

Adherence to the five-phase strategy helps build a lean organization one brick at a time. The strategy provides guidance for determining whether the organization is progressing at an appropriate rate. The five phases are not a rigid system of implementation. Project managers must be flexible when responding to changes during the physical implementation.

Sometimes the path toward lean becomes unclear. When this happens, a few simple questions may get the implementation effort back on track:

- Where in the five phases is the implementation?
- What is the resistance to moving forward? For example, resistance to standardized work usually results from a feeling that it is a way for supervision to catch an employee doing something wrong. The problem is trust, not standardized work.
- What behaviors are exhibited by implementation leaders?
- Where should the implementation be in the five phases, given the elapsed time?
- Is there enough stability to proceed?
- Is uneven implementation caused by lack of effort or knowledge?
- Should the implementation team be left to struggle, or should a leader or consultant direct their efforts?

The answers and subsequent strategies are not cure-alls. Use them based on the situation, the current stage of the five-phase implementation, and leaders' commitment to supporting the actions. Once these factors are considered, and an overall strategy for implementation is established, then the fit of each is apparent.

Creating a Lean Human Resource System 8

By John Allen

There are two types of individuals in a lean manufacturing organization: those who add value to the product and those who support those who add value. Implementation must focus on those who add value. Line workers are the reason the organization exists, so they must be the focal point of the implementation. Today's organizations do not have an abundance of salaried and support personnel to sustain implementation. Therefore, the line worker is needed as a resource to achieve implementation.

All employees must be satisfied with the answers to two questions before lean implementation can proceed—"What is in it for me?" and "Why should I change." *Change* is a difficult process for everyone, especially for someone who has done his or her job the same way for many years. Many might say that if the job was not being done correctly, why did managers allow them to do it the wrong way for so long. Change must happen because going forward is far more positive than maintaining the status quo. When lean proponents say it is the right thing to do, but the rest of the organization is striving to maintain the status quo, the likelihood of a lean reality is greatly reduced. Many have come to believe that the only way to move an organization to this level of change is to have a significant emotional event.

Change is based on logical construction of the current situation and a vision for a better way. It starts with an honest look at the driving force in organizational life. This driving force is commonly referred to as *competition*. Competition can make a company stronger or it can destroy it. Competition is a catalyst for improvements that ultimately make a company more competitive, which provides the ultimate payback of job security. Every employee is concerned with job security, even when financial security is attained. The job is the basic unit of organizational life and provides each employee with a financial return and a social structure to engage in positive interaction. When the job is threatened, the worker moves very quickly to demonstrate that their position is important. For this reason, change for the sake of change is not desirable. To further inflame the problem, manufacturing organizations collectively have made numerous failed attempts at changes in the past, so workers are naturally cynical of future changes. The first sign of change elicits a reaction that prohibits it from progressing naturally.

Once the true nature of lean is understood, most organizational leaders become zealots for change. However, they also tend to become proponents for a pace that is inconsistent with the true nature of change for most individuals. Results of lean changes are expected immediately. Unfortunately, the true rate of change is proportionate to the level of comfort achieved by those most affected by the change. Going too fast, too early, can delay the implementation at best; at worst, it can halt the change effort completely.

Project leaders first must understand lean as a system, then ascribe to a plan in which employees can engage at a pace suited to gaining *commitment*. Commitment is defined as a willingness to suffer some inconvenience to accomplish something worthwhile. Without commitment, the effort to make changes is seen as an inconvenience and employees will be unwilling to make sacrifices. Therefore, fostering commitment from the workers is critical to engaging implementation.

MUTUAL TRUST

The ultimate dichotomy emerges when leaders want performance increases immediately, yet the worker needs time to become committed. How does the lean implementer achieve balance? The answer lies in establishing suitable expectations and gaining *trust*. Leaders are expected to make a small sacrifice in the short-run to gain tremendous returns in the future. Workers are expected to support the change effort, even though they do not clearly see the benefits. Both expectations are linked with trust. If trust is not present, leaders do not see the logic of postponing results, and workers do not buy into the logic of trying something new and uncomfortable.

Trust must be earned over time. Trust results from having a measure of predictability in the relationship. Simply talking about it is insufficient—it must be observed. In a situation where workers do not trust leaders' motives, taking one step at a time becomes critical. Each step becomes a demonstration of how the implementation will proceed. Careful planning and evaluation can ensure that the steps taken are viewed positively. Some steps may result in gains, and some may not. Understanding the overall goal of achieving a lean operation provides leaders with the wisdom to take steps that do not necessarily result in immediate improved performance. Those who understand the ultimate goal enhance the level of trust.

Project leaders must understand that the entire organization, not just the line worker, will test the implementation to see if it is a true way of doing business, or just another "flavor of the month." These tests usually begin simply. A relatively unimportant issue is chosen to be a small measure of dedication to the process. For example, if employees are asked what would make the implementation go better, they might respond with, "softer toilet paper and brighter light bulbs." These are not vital issues to employees, but a test to see if a request that is irritating to leadership is taken seriously. Other tests involve issues such as:

- how overtime is handled;
- how people who have different ideas are treated;
- how job rotation is handled, and
- how the results of the implementation are shared with the workers.

A consistent response from leadership about lean helps build trust. Over time, trust is built one decision at a time until the lean direction is understood and can be committed to by all parties.

As the foundation of trust emerges, it becomes obvious to the worker what is in it for them. The environment is aimed at making them successful. With the changes, they can do their best work and produce the highest quality. They are treated with respect and mutual trust. They are communicated with fully, resulting in their trust and respect.

THE IDEAL LEAN HUMAN RESOURCE SYSTEM

To characterize the ideal lean human resource system, project managers must begin with a vision. In most organizations, lean manufacturing is a new concept and requires a vision that characterizes the desired future state. This vision must be written and communicated by those who will implement it. It must be a clear and compelling departure from business as usual. Leaders must articulate the vision so the organization is encouraged to take action, which will lead to the desired outcomes. For the vision to be understood, the message from leaders must be consistent to gain trust from employees. The vision must project a clear picture of what the organization will look like in the future.

Vision

Once a vision is established, the correct mindset must be developed to drive progress. The proper mindset accepts that problems are welcomed and are good for the organization. In a lean organization, a problem is defined as a deviation from an established standard. Implicit to problem-solving is a decision to make standards and expectations clear and explicit. Problem-solving can then progress as the standards become a basis of comparison for current performance.

Too many organizations have unwritten rules that bringing forth problems is bad for the career of the individual. These organizations put forth the notion that showing problems is a weakness, rather than a strength. Therefore, problems are rarely discussed in a public forum. As a result, opportunities for improvement (problems) are hidden, and the growth of the organization suffers.

Problems are Good

It is important to establish a "problems are good" philosophy among employees. This philosophy requires that lean implementers have solid problem-solving skills. It also requires a conscious, positive effort to recognize those who bring forth problems. Then, implementers must ensure that the person elevating the problem is the one solving it. In this way, problem-solving becomes the value, not just problem identification. Elevating and solving problems are qualities that should reside in all employees.

The production system requires appropriate responses to whatever questions or issues arise. These include quality problems, a breakdown of the standardized work process, equipment breakdowns, and many other issues. As these situations are discussed in a group of peers, roles should be established to ensure the appropriate response is made. For example, an operator signals that a problem exists in the completion of his standardized work. If roles are aligned properly, a team leader or supervisor arrives in the workstation and assesses what help is needed. As part of their role, supervisors know they must manage the contradiction of keeping the line moving and delivering the highest quality from the workstation. The line may have to be stopped while the quality problem is fixed. If roles are not clear, the supervisor may simply decide to ship it, thereby detracting from the highest quality.

Monitoring the System

An ideal human system has mechanisms in place to constantly monitor employees' morale. Data on employee morale can help determine where deviations from the expectations exist. The data is usually compiled from three sources:

- opinion surveys;
- daily interaction between workers and leaders, and
- constant walking of the shop floor by employee relations representatives or union representatives.

For these processes to be credible, problems must be acted upon in a way that supports the principles of lean. Trust is a natural outgrowth of the monitoring processes. Trust comes from consistent application of policies and procedures. Employees within an organization base their opinions of fairness on how simple problems are handled. Human resources personnel are in constant contact with the situation on the plant floor. Constant contact enables operators to feel represented, which builds credibility through the process of solving problems. Representation becomes a check and balance for the health of the people system.

The human resource representative also monitors the organization to determine if the company is developing *citizens*. A *citizen* is a member of the company who is able to make informed choices that benefit growth. For a person to be a citizen of the company, he or she must fully communicate the goals and objectives of the company, as well as the current situation. Communication becomes a business objective to help develop the company citizen.

Enlightened Leadership

Leaders and managers must be enlightened to lean concepts and principles in an ideal lean people system. Enlightenment comes from understanding the vision, and making decisions consistent with that understanding. The process of preparing leadership for the responsibility of leading the lean implementation requires training and exposure to successful implementation programs. Leaders must be able to take advice from experts. The successful preparation of the leaders for this responsibility requires time and should not be skipped. Many times the outcome of the lean implementation depends on the quality of preparation.

Before leaders can be successful, they must define their roles, which is a difficult process and requires substantial time to accomplish. Care must be taken to not define roles so tightly that people feel they have a clearly defined box around their responsibilities. At the same time, the defined roles must enhance a sense of teamwork.

Job Security

Job security is not a promise of a job for life. Job security means that as long as the company is making a profit, employees engaged in lean improvements do not lose their job. The worker understands that if the company is losing business or money, certain actions to reduce cost are necessary. If these actions are not taken, then the fundamental contract with the employee—payment for work accomplished—cannot be fulfilled.

Often, the connection between job security and actions the worker can take to ensure competitiveness is missed. By feeling secure in a job, the worker is free to make improvements, including the elimination of their job. They know the company will place them in another position that is at least as good as the one eliminated. That may include a position on a Kaizen team, a job involving cross-functional problem investigation, or another equivalent position.

Many companies have gone even further to recognize the priority of actions taken when the economic viability of the business is threatened by creating a list.

1. Increase cost-reduction activities.
2. Eliminate overtime.
3. Bring services purchased outside in-house.
4. Eliminate bonuses.
5. Reduce salaries.
6. Lay off support personnel.
7. Lay off hourly personnel who add value.

Making this priority list known helps establish the importance of the value-adder and gives the worker a true sense of the degree to which their job is protected.

Training

Training is always a problem for plant leaders. Everyone recognizes the importance of training, but it never seems to get done unless the organization embraces a simple principle: growth depends on developing personnel. Until this is recognized, training will never be seen as a priority. There is always too much pressure to produce.

Training must be aimed at building worker competency. However, worker development needs to be a good investment. Training for the sake of filling seats in a classroom becomes entertainment. It must develop competency on the job. Competencies are behaviors that lead to success on the job. It is the collection of behaviors that, when performed, creates successful outcomes on the job.

The Rational Way

If a manager is focused on the worker who adds value, he or she quickly realizes that every method, process, and procedure must be easily understood. Therefore, choosing the rational way for interactions is an important element of a lean human system. This element ensures that information and solutions passed downward through an organization are easily understood and acted upon. Many times, managers tend to create solutions that are driven by engineering. These solutions are generally difficult to understand and require an engineer to implement. People who add value should be able to easily understand and implement them. This takes more time, but it is time well-spent toward empowering the line worker. In addition, rational solutions should be less threatening and readily maintained. These characteristics are an outgrowth of the improvement accomplished by line workers, and have direct implications on how the rest of the organization provides help to the line worker. Making solutions simple and easily understood is the aim of every interaction. Having the entire line work force enlisted in the battle to remove waste is necessary, and keeping change at a level that encourages simple solutions is imperative.

Manpower Planning

The ideal lean people system must balance long-term thinking and short-term action. No other element better exemplifies this than manpower planning. Without a long-term plan for the number and type of people involved in the organization, the likelihood of appropriate short-term hiring and training is lessened.

As with other aspects of the production system, manpower must have a long-term objective. In the case of manpower, that objective is to protect job security. Job security is protected when short-term hiring achieves the near-term goals, while prohibiting the longer-term problem of having too many people to be competitive. Several choices exist to handle short-term hiring needs, including temporary transfers, contract hiring, and temporary services. Each of these gives the organization an opportunity to reduce head count, yet not impact job security and the relationship with regular employees.

Recognition

When employees are asked what characteristic they would like to see in their organization, the reply always includes more recognition. Honest recognition is the purest form of information about job performance, and can be used to shape future performance and motivation. In a lean organization, recognition points the way through the ambiguity of implementation. Many aspects of lean appear to be counter-intuitive, such as producing less when you can produce more, constantly searching for problems, or producing at the slowest rate that satisfies the customer's demand.

Supervisors and managers give recognition to those who exhibit successful behaviors. The behavioral aspect of performance is a critical linkage because, in the short term during the stability phase, there may not be measurable results. Behaviors that lead to favorable business outcomes need to be recognized consistently during the implementation.

ESTABLISHING THE GROUND RULES FOR THE RELATIONSHIP

The most important consideration in creating a lean people system is to ensure that relationships with employees get off to the best start. These relationships are based on the same principles that guide the lean people system implementation because the production system is the culture. Expectations for both the system and the people are set by the production system. Therefore, any effort to create a culture distinct from the production system is counterproductive to the overall changes. For example, in many cases organizations want teams immediately. However, the real need for teams arises only when the first time line re-balance is required during the standardized work phase because of changes in customer demand. By promoting teams too early, the lessons of maintenance of standard work get lost because they only require individual action.

The expectations set for employees need to be congruent with the mission, values, and guiding principles of the organization. The mission is distinct for each organization, and gives broad guidance as to how the organization will achieve its values and principles. The values and principles should include statements like the following:

- Always strive for cost-consciousness.
- Problem solutions must be rational to the shop floor.

- Strive for mutual trust and respect between employees and management.
- Put the customer first, and maintain competition and cooperation with the industry.
- Have respect for the value of people and promote challenge and courage.

The written word is important, but demonstrating basic values and principles is critical. Leaders of a lean people system supporting a lean production system must take care in giving meaning to these values and principles. They become part of the ground rules and expectations for employees as they mentally match what they are hearing with what they are seeing.

Selecting and Assimilating Employees

Another powerful factor in establishing the ground rules for behavior comes through the hiring process. The selection system can send some very direct signals to potential employees as to what is expected. Many organizations perform "day-in-the-life" assessments to observe the candidate working in a situation that approximates the real work environment. This allows the employer to observe behavior and communicate what is expected on the job.

Those who successfully pass the assessment are given structured interviews that once again send straightforward signals about important aspects of the job. The combination of behavioral observations and behavioral samples allows the company to select the best fit. In addition, observations and samples provide candidates with specific information about expectations so they can make a good choice of employment. Thus, the formation of ground rules occurs before the person is selected.

After the employee is selected, ground rules are further enhanced in the assimilation process through interaction with company managers. Managers should make an effort to see that these interactions are consistent with the behavioral expectations of the lean culture. Inconsistencies can result in a lack of trust by employees and difficulty in getting them to accept new processes. There are several key opportunities for managers and implementers to set the appropriate expectations, including:

- the role of the employee relations organization;
- the mindset of the leadership;
- to first involve and then empower employees;
- the overall organizational structure, safety, health, and ergonomics focus, and
- the use of job rotation to emphasize flexibility in the work force.

Employee Relations

The employee relations (ER) organization is challenged with the task of maintaining consistency in decision-making. It serves as the conscience of the organization. Its responsibilities include communications, discipline, recognition, employee problem-solving, supervisory relations, attendance, and policy development. These responsibilities are opportunities to either reinforce the culture or fragment it. Early in the introduction of the lean enterprise, employees are looking for clear, consistent messages supported by actions. Inconsistencies may confuse and discourage employees, who then do things to protect themselves. It is important to understand this if the company intends to make rapid changes to the operating system. Without a clear message, workers will be reluctant to aggressively implement the lean production system. The cost of going forward becomes much greater than the cost of maintaining the status quo.

The ER organization has the responsibility to create, monitor, and interpret the company's policies. A policy in a lean organization creates the expectation of how someone will be treated under prescribed circumstances. These policies should be prominent when interacting with the employee. Consistent policy interpretation is measured by using the same factors each time a decision is made, not by making the same decision each time. Ensure clear communication with the employee so they see an interpretation as an extension of the expectations previously established.

Leadership Mindset

To create the behaviors necessary for implementing lean, company leaders must establish a clear mindset that is consistent with the values and principles. This mindset includes the belief that employees can contribute to the company's success. The mindset also must include:

- a willingness to explore the potential of the employees;
- a customer rather than a power relationship with the employees;
- a willingness to shift power to the line worker;
- a willingness to provide support, and
- assurances that everyone supports the preceding items and they are willing to eliminate their own job if necessary.

This leadership mindset joins a work site management style with several discrete facets. Work site managers must encourage everyone's self-development. They must listen and respond to employee concerns. In addition, they must be an example by following work rules and regulations, and respond to infringements. In doing these things, they lead the work force by example.

Organizational Structure

Lean implementers will find it necessary to examine the way the organization is structured. The organizational structure is established for one reason, to support the line worker. Support means that any problem occurring on the line is given immediate priority so the work process can be completed within Takt time. This support requires a method for signaling help, providing control for the work group leader to respond to issues in a timely manner, and enhancing problem-solving skills.

The work group leader's response time to occurring problems is critical. Response time is characterized as the ability to solve problems within Takt time. Usually, the ratio of leader to worker is around five to one on a lean assembly line. This span of support is usually determined by distance to the points where aid is needed, and the ability of the lead person to quickly solve problems. Once the organizational structure is established, consistent monitoring is required to ensure the line is not stopping too often because the lead person is not responding.

For a machining line, the organizational structure is different because the number of machines is greater than the number of line workers. Issue priority is established with an electronic signal sent by the machine and visually displayed in the machine cell. The lead person monitors the visual display and instructs line workers as to the priority issues. He or she then monitors the quality and timeliness of the response to determine where training and assistance are needed.

Job Rotation

Job rotation is important for initial lean implementation, as well as the long-term viability of the lean enterprise. In the short run, it helps prevent work-related injuries due to repeated movement. It also keeps the worker excited about learning new responsibilities. Job rotation should become routine in the stability phase. It requires support from the lead person to train the line worker in their new responsibilities. Training should take the form of a simple, repeatable process with the line worker as the customer.

Job instruction training is the process of training someone in a new job. It is simple and based on the assumption that unless a student learns, the teacher has not taught. Making the leader of the work group competent in job instruction training is a key to having the line worker accept job rotation. If the work group leader is not a competent teacher, the risk is too great for the line worker to rotate and they will invoke seniority to remain at their present job.

MAINTAINING THE ESTABLISHED RELATIONSHIP

Once the ground rules are established, project managers must shift their focus to maintaining the gains achieved in the implementation. Maintaining expectations is by far the most difficult process when implementing a lean people system. It requires constant attention. This maintenance is not a random process, but a series of specific interventions that keeps the momentum going.

Policy Deployment

The key to maintaining expectations lies in the process of goal alignment and commitment. This process is known as *policy deployment*. Policy deployment begins with an honest assessment of the operation's current state, and continues in ongoing cycles that never culminate. The steps to policy deployment include writing a long-term plan for business success. This plan progresses through goal-setting in different layers of the organization to support long-term and annual goals. At each step, policy deployment makes sure that plans and goals are aligned, and that those responsible have the resources to complete the objectives.

Once goals are aligned and have been committed, supervisors monitor performance to achieve the goals. Monitoring takes the form of either coaching or formal reviews. In either case, the aim is to ensure that performance matches the expectations established in the planning process. The quality of the supervisor's annual and long-term goals determines whether they must be reconfigured to reflect the current situation. Changing them should be a rare occurrence.

Information gathered in the final annual review becomes a part of the planning process for the following year. With this information, the cycle is both completed and started again at the same time. Data gathered in the annual review process influences the annual and long-term goals for the following year.

The essence of policy deployment is in the quality of discussions within the organization. The quality of discussion in an alignment meeting relates directly to the level of the deployment plan and the full understanding achieved by the plan developer. A discussion has poor quality if a manager allows a plan to proceed without believing that it is understood and supports the goals of his or her area. If the manager knows the goals are understood, and the plan is written in such a manner that appropriate action ensues, then the quality of the discussion is good.

Another measure of quality in the policy deployment process is the extent that the human and material resources matches those needed to accomplish the goals. It may be in the manager's best interest to see that resources are scarce. However, if they are so scarce that the likelihood of failure is great, then the manager has failed the process.

Policy deployment is the basis for maintaining expectations. There are other supportive methods available to ensure that expectations are maintained.

- Create a recognition and rewards program for showing that successful behavior is essential.
- It is helpful to have a consistent, disciplined problem-solving process.
- Opinion surveys are useful for understanding morale and support issues.
- A suggestion system that supports small gains made in the implementation is useful.
- It is necessary to have supervisors managing daily performance.

Manpower Planning

The next major consideration in maintaining the relationship among stakeholders is to put a viable manpower planning process in place. This process is critical to having the right amount of people in place and establishing a system that ensures job security. Manpower planning defines the flow of people coming and going from the facility.

To begin manpower planning, a project manager must have a good baseline of people by both category and department. The common categories used are:

- direct labor personnel, including team leaders who spend an average of 50% of their time working on the line;
- those who directly support the line, such as conveyance, maintenance, quality, and production control personnel who are directly associated with the line;
- all salaried personnel not described previously, but not management, and
- all management personnel.

This breakdown helps allocate personnel properly and maintains a balance ratio of four direct employees to every one indirect.

Planning should project outward at least three years, and an accounting should be conducted for each major event that takes place. Events can include personnel turnover, additional required personnel, expected promotions, major shifts in responsibilities, and transfers from one responsibility to another. A definitive plan for manpower needs can then be created that describes the number of people by department and by classification. From this information, a strategy can be developed that enables human resources personnel to manage hiring, transfer, and promotion issues to protect job security.

The issue of job security is so pervasive that once workers understand the manpower plan, they will assist ER in maintaining it. This is unlike the usual relationship that exists because production departments routinely fight for more people. One issue often misunderstood is what impact current decisions to add people have on future job security. Achieving consensus on the manpower plan helps create an understanding that allows ER to effectively manage manpower numbers.

Leading By Example

No issue is more important to maintaining relationships between workers and leaders than the examples set every day by company leaders. Leaders are tested every day. They must successfully pass the tests to ensure that the proper messages get communicated to the organization. These messages are about what is important and how a predictable relationship can be created and maintained.

Leaders must answer fundamental questions each day about how the organization is going to proceed. The most important question is "what are the real standards for the organization?" The answer in a lean facility lies in the way workers add value to the product, and the way they are supported in the process of adding value. For the worker adding value, standardized work sets work methods and maintains them over time. Leaders must insist that standardized work procedures are followed. They know the significance of drifting from the established standard. Drift causes the organization to lose efficiency and achieve far less than is possible. Leaders must observe how people perform to their standardized work each day and not allow deviations to occur. Problems must be solved to eliminate deviations and move performance back to the standard.

Once the principle that those adding value are the most important is realized, leaders must constantly be asked by these people "how important am I?" This test represents a desire by value-adders to never let the organization forget their importance. The context for the question centers around issues of absence, pay, benefits, disagreement with supervision, disagreement with fellow workers, and policy interpretations. Each time a question about these issues arises, leaders must understand that the issue is more than the presenting problem. Rather, it is an opportunity for leaders to reinforce the agreement established in the policies, principles, and values espoused.

Involve the Value-adders

Leaders must directly involve those who add value in designing the workplace, and make sure they are rotated to achieve flexibility and desirable ergonomic practices. The training area is another opportunity in which to show importance. Insisting that people be trained in leadership establishes a pattern for enhancing the person and is a basic strategy for growing the organization.

Most perceptions of importance come from the communication pattern. The frequency, level, and attention given communication become a practical demonstration of its importance. Leaders and lean implementers must make sure that communication flows freely in every direction.

Periodically, issues such as overtime arise to test the leadership. Hourly paid personnel want additional income, but they also want time away from work. Even the most zealous advocates of working overtime will find a way to take time off to compensate for additional work. In a lean organization, overtime is only worked on a daily basis to compensate for not achieving the daily schedule. One exception, however, is when short-term commitments for product overshoot the availability of hours used to calculate Takt time.

In many operations, overtime has become a disease that compels employees to work to maintain a standard of living they achieved during prolonged periods of past overtime. These employees do whatever it takes to maintain that standard of living. Expectations have been set and they are not easily changed. This issue becomes a significant challenge to an organization's leadership. Managers must react appropriately to maintain the daily schedule, while not forcing personnel to take actions that make overtime required. In these circumstances, attaining the daily schedule can only be maintained if standardized work is established. The best way to achieve this is to have overtime available only for improvement activities. Personnel can work overtime while standardized work to a Takt time is preserved.

Leadership Mindset

Changing to lean must begin with a change in the mindset of leaders. Often, this change is a clear departure from the traditional organization, which requires management to view itself as distinct from the process. In a traditional situation, managers do not add value. A traditional situation is usually characterized by feeling a need for separation between thinking about business and doing value-added work. Employees are expected to operate as individual entities that make as much product as possible. Any improvement in their work life comes from managers making decisions that affect them. Traditional organizations believe that managers are best prepared to make decisions by virtue of their education, or because of the social levels they have attained.

Working in a lean enterprise requires a completely opposite mindset—to think of the facility as a functionally interdependent unit. The unit must consider the impact that decisions for one part of the organization have on the rest of the organization. For example, what does the impact of reducing inventory in one segment have upon the flow of the rest of the organization? The organization must change from a collection of lone experts to groups of collaborative teams. The transition must eliminate functional boundaries in favor of a system-wide perspective.

Embedded in the thought process of a lean organization is the belief that everyone in the organization is capable of contributing. This often leads to the resolution that each individual is worthy of being developed. Time for training becomes a business strategy, not an inconvenience. Training time is subtracted from the available time when calculating Takt time. Training reliability becomes a focus of leadership's attention, and gaining competency becomes a necessity. On-the-job behavior consistent with training objectives becomes a driving force for reinforcement. Supporting supervisors and team leaders must be prepared to monitor behavior and provide coaching to support the training.

Leaders' mindset has to shift from impulsive decision-making to disciplined problem-solving and implementing countermeasures. This shift necessitates having problem-solving skills and an insistence from leadership to follow established processes. The organization must eliminate the old habits of solving problems on the basis of opinion as the only method.

For the line worker, complete understanding of job security is the basis of moving forward. Line workers must see themselves as the protectors of job security. They must understand basic business principles to help support decision-making and challenge inconsistent conclusions. Line workers must be committed to disciplined action so that standardized work can become a reality. They must be willing to tell leaders when they make wrong conclusions. Likewise, management must be willing to listen.

MANAGING THE PEOPLE SYSTEM

Managing the people system is a difficult process and there is no formula. Every interaction with the employees of a lean organization is an opportunity to either remove or destroy waste. Ensure that actions involving the people in the organization are thoughtfully considered. Even the simplest interactions, such as written communications, should be planned and be consistent with enhancing the theme undertaken at the time.

Clearly, leadership is the key to the pace and depth of the implementation. Leaders must realize that lean implementation is a "contact sport." Going and seeing the actual situation cannot be an afterthought; it must be the normal part of every day's routine. Being in the area where value is added enables leaders to see what is actually happening, and allows them to measure performance against standards created in the policy deployment process. Once discrepancies are found, a leader can use the skill of matching his or her leadership style to the situation.

- In places where workers do not have the skills or motivation, a directive style may be necessary.
- In situations where workers have some skills or motivation, a coaching style can be used.

- In situations where workers have skill but lack motivation, a participative style can be employed.
- In situations where followers have the skill and the motivation, a delegating style can be used.

In all of these situations, the leader is matching the style of leadership with the development level of the followers.

Planning

Having a simple plan for leading employees in a lean implementation is critical. Too often, planning is delayed for the sake of getting to the action (the implementation). For a leader, the plan is a concrete measure of employees' development levels. Plans show the thought process, level of attainment expected, and level of resources required. Using this information, the leader determines how much assistance the employee needs to achieve the established goals. The form the plan takes is highly variable and can be a source of information about the leader's ability to organize his or her thoughts.

Planning becomes the key process in the first step toward implementation. The first challenge for the leader is to insist that there is enough time to ensure successful planning. It is a test of the leader's commitment to a smooth implementation. Even though the workers implementing lean may want to proceed immediately, they must master the plan before they proceed.

Resistance to planning is a common occurrence. This resistance can be overcome by making the planning process fun. Leaders can gather together those involved in a room with lots of sticky notes, then lead the group through a planning exercise. Ask the group to write as many tasks as necessary to accomplish the goal and stick them to a large wall. Discard duplications and irrelevant items so that only the tasks necessary to complete the goal remain. The tasks are then grouped and placed on the wall in chronological order. As they are put on the wall it becomes apparent that there are gaps in the plan. List the gaps and place them in the proper sequence.

After the initial plan is created, having a discussion about how far and how fast (a time scale for implementation) provides information about start and end dates. From this information, a productive discussion about required resources can be conducted. A simple plan has been created with the information necessary to then use a computerized project management tool. This document then becomes the tool the leader uses to manage the project. In addition, the plan can serve as a guide to developing subsequent plans.

CONCLUSION

Clearly, setting the appropriate ground rules is crucial to implementing lean. Ambiguity in the company's expectations results in a lack of trust within the work force. Without trust, the line worker is inclined to remain in the current system, even if it is clearly inferior to lean. Supervisors and managers must make sure that lean expectations are pursued with diligent monitoring so that the expectations can be attained. Having these expectations set early indicates that what follows, at a minimum, is consistent with the communication around lean implementation.

Unfortunately, the attention given to creating the human system organization in a lean implementation often falls short of what is required to sustain performance gains made through technical improvements. This is a lesson that apparently must be learned over and over again, as companies discover they really need to invest in their human resources to create lasting change. With proper planning in the initial stages of a lean implementation, however, both technical and human improvements to the organization can provide lasting results.

Section III:
Lean Tools

Policy Deployment   9

By David Stewart

Policy deployment is a process for ensuring that an entire organization is aligned to accomplish lean objectives. It is a strategy to accomplish the organization's top priorities and set annual goals. The policy deployment process focuses on systems that need improvement to achieve strategic objectives. The key systems typically remain the same year to year, while goals are adjusted to provide a higher level of performance.

Policy deployment is standardized in that it enables all the facets of an organization to act in unison to achieve company-wide goals in strategic areas. The ultimate objective of policy deployment is continuous improvement. The Japanese term for policy deployment is *hoshin kanri*, which is defined as a planning method for setting strategic direction. Originally, hoshin planning was a form of strategic planning designed to accomplish several objectives that are critical elements of the policy deployment process:

- Identify important opportunities for improvement.
- Determine the most effective actions to accomplish the improvements.
- Create a detailed plan to implement the actions.
- Provide a means for reviewing and correcting the plan and to retain lessons learned.

"Deployment" is a military term that means "to organize troops or equipment so that they are in the most effective position." Similarly, policy deployment means organizing a company for effectiveness. Without a standardized process by which to coordinate activities, companies can go in many different directions resulting in duplication, conflicting goals, and poor results.

PRINCIPLES

Traditionally, companies have always had some form of policy deployment. However, policies are often created by high-level managers and mandated to the rest of the company, without sufficient empowerment or resources given to employees to accomplish the desired results. These business plans typically are disconnected from the rest of the organization, and there is little involvement from those who have the most influence on achieving the business plan—the workers.

When planning a lean implementation, it is important to begin developing the policy deployment process before any new systems are implemented. Policy deployment becomes a part of the change management strategy to facilitate communication and coordination of effort. Discipline is important when negotiating goals and objectives for policy deployment so the desired relationship between supervisor and subordinates can be established. The dialogue that occurs during this process becomes an ideal opportunity to develop subordinates and help them understand how their efforts benefit the company.

Policy deployment is well-suited to the task of communicating the future state vision and objectives toward the overall lean vision. Employee buy-in during deployment creates an alignment that is necessary for negotiating the difficult waters of change. Continuing to develop the policy deployment process after implementation serves well in the quest for continuous improvement.

One goal of policy deployment is to achieve horizontal and vertical alignment between all parts of the organization so everyone is efficiently working toward the same goals. Alignment within an organization is especially important during times of change, when people are often asked to carry on their own work as well as implement new processes and procedures. Not only will policy deployment facilitate the transition to lean systems and procedures, it will provide a means toward the culture of continuous improvement.

Roles for Personnel

Policy deployment does not only apply to production departments. The entire organization must be involved, from manufacturing to engineering to human resources, production control, and finance. Manufacturing departments define their objectives and work with support functions (for example, quality control and materials) so their activities align with the requirements of manufacturing. All levels of the organization must participate and coordinate the planning, development, and deployment of annual objectives. For example:

- Production operators agree to strive to reach the targets and acknowledge the reality that their success is directly tied to the success of the company.
- Production supervisors communicate openly to subordinates, explaining the key systems of safety, quality, cost, responsiveness, and organizational development at the beginning of the annual planning/policy deployment process. Supervisors are accountable for a portion of the annual planning/policy deployment, and for managing their plan by relating it to their subordinates and making it a part of the operators' annual objectives.
- The staff level and support sections of the company have their own accountabilities, which tie directly into the annual planning/policy deployment process.
- Middle management is responsible for ensuring that policy deployment objectives are met and directly correlated with their subordinates' actions and efforts for achieving objectives.

By taking ownership, everyone within the organization understands that if the company is successful, they will be successful as well. The agreed-upon goals and action plans that cascade through the organization must be based upon the true capability of the organization. These goals are then reflected in each person's annual performance review.

ADVANTAGES

Policy deployment provides a means for coordinating activities so that each effort is aligned at all levels of the organization. This ensures a systematic method of providing a central focus and direction for everyone's efforts. It also establishes quantifiable targets and goals. If these targets and goals are not clearly defined, the efforts and effectiveness are impossible to measure.

Leaders and subordinates both must develop action plans that meet company and personal development objectives. Policy deployment helps establish this by challenging managers and subordinates to take ownership in creating systems and methods to achieve objectives. They also must facilitate the planning to manage these objectives.

Policy deployment provides a consistent method for all personnel to measure the performance of their area, their subordinates, and themselves. It is unfair to all persons involved if they are not aware of the standards upon which they are evaluated. With clearly defined goals and expectations, it is easier to highlight and recognize improvement opportunities. This stage is crucial in standardizing the policy deployment process. Policy deployment helps ensure that all departments and processes of the business are aligned and are actively pursuing the same desired results.

Policy deployment is a methodical approach for planning and assigning activities to employees to increase their opportunities for personal development and advancement. The desired outcome of the deployment process is establishing the focus and driving the initiative of individual efforts toward continuous improvement. Aligning business objectives and personal activities with agreed-upon objectives supports an equal opportunity to make a difference in the future of the employee and the company.

PROCESS

Policy deployment involves a sequence of steps taken at different levels of the organization. Top leaders initiate the planning process, but it is not committed to until input has been received from all levels. Plans are executed with a coordinated effort by all employees to achieve the annual objectives. Progress reviews are critical to ensuring that the organization remains aligned and that progress is being made. Each of these steps is examined in more detail in Figure 9-1.

Planning

The planning phase in the policy deployment process comprises three steps: establishing a vision, formulating a three- to five-year

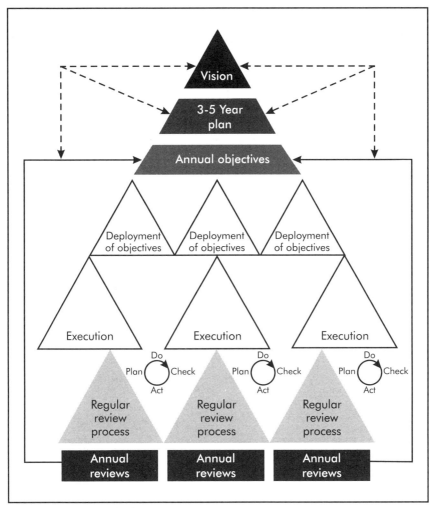

Figure 9-1. Policy deployment steps.

plan, and setting annual objectives. Planning begins by first looking toward long-term objectives, then moving systematically to examine mid-term goals and annual objectives. Leadership creates a long-term vision, then develops a three- to five-year plan. Annual goals that support the mid-term goals are selected, and these objectives are then rolled out to all departments. However, the first step in the planning process is to go back to basics and examine the organizational vision.

Vision

To ensure that the planning process is consistent with the desires of the organization, it is important to start planning by creating or revisiting the organization's vision. This vision is developed by top leaders with a long view to a future that is currently out of reach, but potentially within grasp. The vision should be compelling and consistent with the organization's mission and purpose. The vision establishes the direction of the organization's future movement.

Once the top managers complete a draft of the vision, it is reviewed by the rest of the organization. A final version is created that incorporates the input from various levels of the organization. The final vision is then clearly communicated to everyone at all levels.

Develop Three- to Five-year Plan

Once the visioning process is complete, the leadership selects vital breakthrough objectives. Breakthrough objectives should be distinguished from incremental improvements. The vision describes long-term challenges; breakthroughs identify mid-term opportunities; and annual objectives specify short-term tasks. Breakthrough strategies are a bridge between the challenge of the vision and practical and tangible goals. This mid-term plan maps a three- to five-year journey toward the vision.

To design the plan, improvement opportunities must be identified, which are determined by focusing on the gap between the organization's present capability and the performance specified by the vision. Data from the corporation's past performance, present environment, and competitors must be accurate to determine the present capability of the organization. Once the objectives have been identified and the three- to five-year plan is developed, it is communicated clearly to everyone at all levels, just as the organization vision was communicated.

Develop Annual Objectives

Annual planning objectives are developed from the three- to five-year plan. Annual goals are integrated parts of the mid-term objectives, and consist of a small number of focus points.

Policy Deployment

Developing annual objectives requires an accurate assessment of the previous year's performance as it relates to the top priorities for the following year. Make sure the data used when selecting one-year objectives is accurate. As leaders develop annual objectives, feedback from their subordinates is requested. As feedback is integrated into modified plans, those who had input are asked for their willingness to support the stated goals.

The reason for improvement must be compelling to all levels of the organization. It is insufficient to say that everyone needs to improve. It is important to explain *why* it is necessary. Everyone wants to know the reasons and importance of the decisions made and their part in the plan to achieve improvement. Explaining its necessity facilitates buy-in and commitment. This supports the concept of "when I am successful, the company is successful, and when the company profits, I profit."

Typically, the plant manager develops a draft that is reviewed by department managers. Managers' feedback helps shape the final objectives. The managers are then asked to support the annual objectives. When there is consensus, the objectives are communicated to the entire organization.

Roll Out to Departments

Following the development of annual objectives, the plan is rolled out to departments to develop the targets and means for reaching the objectives. This involves asking the employees who will be responsible for implementing the plan to design the plan. It is crucial that senior managers use consensus and team reviews during this process, not one-way directives. Employee participation then becomes a means to clarify understanding, build consensus, and gain commitment to action. This process is called "catchball."

Catchball is a term used in policy deployment to refer to iterative planning sessions that take place during the negotiation process. Catchball is a metaphor that describes a planning process in which objectives and plans are tossed back and forth between levels of an organization until consensus is reached. The catchball process encourages employees and managers to talk with each other until they reach consensus. Agreement often takes more than one or two planning sessions. Catchball creates constant clarification so that alignment is the result.

The objectives for a particular area are aligned vertically and horizontally. Senior managers cascade the annual objectives, and as they travel down the organizational hierarchy they are more specifically defined in terms of that area or function. Employees negotiate the means and measures with peers and managers. Managers at each level consolidate the plans for their particular areas to verify that they meet the stated objectives.

When employees are told what the big picture is, they can identify their role in achieving the objectives. During the cascade, the means of the level above become the objectives for the next level. For example, if supervisors agree that improving quality is the means for reducing repairs, then the objective for the operators is to improve quality. The means that supervisors develop for achieving the objective becomes the objective for the level below. The individual closest to the work describes the most appropriate strategies to support the objectives. In other words, one level specifies "what," the next layer specifies "how." It is important to distinguish measures from goals, as measures are the means by which one determines whether or not a goal is reached.

Planning is only half over when the objectives and means are cascaded from top to bottom. When the objectives have been translated into local means and measures, the roll-up begins. Roll-up provides bottom-up feedback to the leadership on the objectives' feasibility. All managers are responsible for reconciling the plans between the groups below them. They must also verify that the plans meet requirements specified by the next level above. The roll-up is essentially a capability check. Does the collective effort meet the annual requirements? Dialogue during roll-up closes the gap between the business requirements and organizational capability. Responsibility matrices are shared among everyone involved in the deployment to identify conflicts or redundancies. Consolidated matrices that summarize responsibilities are then created, which formalize the modified plan.

Communicating final plans clearly and concisely to other levels in the organization relies on an accurate understanding of all the elements. Once the objectives for each level and department are clear, each level cascades the information to the next. The top level defines and describes it to the department heads, including the portion of the objective or goal for which each department is responsible. Department general managers then communicate the objective to the section managers and assign each of them a portion of that objective. The section manager then provides the same information to his or her assistant managers, along with the part of the mission for which they will be held accountable. Each assistant manager then does the same for his or her group leaders. Group leaders in return comply and follow through with team leaders and down to the team member or operator.

The following is a sample process between a supervisor and his or her subordinates. The supervisor describes what needs to be done with a few annual objectives. The "what" is presented with the "why" (the compelling reason for change). Subordinates detail how the objective will get done. This process is performed by a team manager who completes a standard form with goals, performance measures, and milestones identified. The supervisor then follows a series of steps, including:

- reviewing the report for completeness, realism, and an assessment of needed support;
- assessing the likelihood of success of the objectives based on the competence of the reporting personnel;
- gaining commitment based on an open and honest dialogue and, if necessary, a reworked plan;
- providing the needed resources to accomplish the goals, and
- authorizing the plan to be implemented.

The next level of personnel then repeats the process with his or her subordinates. (See Figure 9-2.)

IMPLEMENTATION

Implementation follows the lengthy planning process. Each goal is implemented from the plan, and performance is monitored daily by a supervisor. Detailed action plans are necessary to manage the projects involved. The implementation plan becomes an active management tool. It translates the annual deployment plan into daily activities.

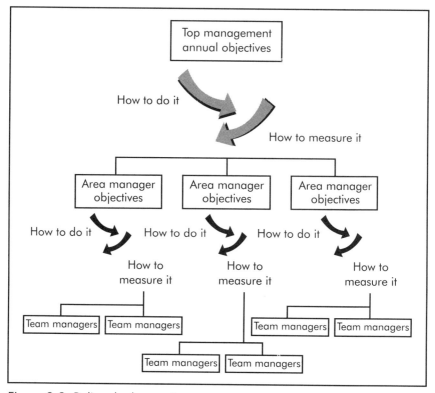

Figure 9-2. Policy deployment process.

Deviations should be discussed as they happen, and changes to the plan should be made only with full agreement by everyone involved. New standards should be documented and applied in other relevant areas. The plan-do-check-act (PDCA) management process can be used to deal with problems that emerge during implementation. Supervisors support problem-solving by coaching and providing resources.

Policy deployment is a visible process. Annual goals and targets for the company, department, and area that are agreed to, established, and clearly communicated, allow recognition and reinforcement in real time. This is critical for all levels of every department to reduce the variety of outcomes, as well as standardize the criteria against which every individual is measured. When efforts to achieve the desired targets begin to fall short, the problem is evident and visible. This serves as a "visual factory" element, and assists managers to put support and assistance where it is needed the most. The theory is to work within the designed system of lean and recognize opportunities for improvement. It is better to have downfalls, recognize them, and use the time to practice the theories than to ignore abnormalities and allow them to fester or get worse.

Regular Progress Reviews

Progress must be reviewed frequently. Regular reviews encourage adherence to the plans, reporting progress toward goals, identifying obstacles to progress, and timely corrective actions. Reviews start at the project level and roll up to the corporate level. A monthly review identifies obstacles to progress and evaluates the need for early corrective action. A quarterly review consolidates individual progress reports to assess their collective impact on the overall policy deployment strategy.

The review process emphasizes that the team diagnoses its targets and processes. Problems are identified, causes determined, and corrective actions planned. A standardized review format and language based on the PDCA management model should be used throughout the organization. The PDCA problem-solving process is used throughout the review process.

Problems are opportunities to be surfaced, not skeletons to be buried. When a problem is discovered, reporting the corrective actions to be taken shifts the focus from the check phase to the act phase of the PDCA cycle. The review environment is one of joint inquiry based on facts and analysis, not blame. Emphasis is on recognition, support, and corrective action rather than punishment.

Annual Review

There are a number of goals for the annual review process, including:

- understanding the gap between the expected results and the actual progress to date;

- summarizing unresolved or emergent issues that may affect the next year's plan, and
- documenting and standardizing what went right, and analyzing and discovering the root cause of discrepancies.

Supervisors at all levels review the results achieved and compare them with targets to determine the implications for next year's plan. Data collected during the year and the more frequent regular reviews provide a wealth of information on missed targets and poor processes. Implications are created for changes to next year's plan. All plans are examined, even when the target is met. Recognition is given for accomplishments, not just criticism for missed opportunities. The policy deployment planning process itself is examined to discern ways to improve it.

DEPLOYMENT CONSIDERATIONS

Remember the following points when developing a plan for policy deployment for an organization. Although not a complete checklist, these points, if followed, are helpful in the journey toward alignment of the organization.

- Establish a good, clear vision for the deployment plan.
- Develop supporting objectives for the plan.
- Align objectives to support corporate goals.
- Establish the main points within the plan.
- Decide methods for each activity, as well as a time frame.
- Set realistic and concrete goals.
- Communicate and get buy-in from all those involved.
- Cascade the plan from the top.
- Negotiate targets to support the goals.
- Establish cross-functional support.
- Execute the plan according to the schedule.
- Closely monitor results and progress.
- Adjust activities to stay on track.
- Standardize the activities that have had positive effects on the production situation.

CONCLUSION

Policy deployment requires a significant investment of time and effort to realize its benefits, but it is worth it. In the process of establishing and developing a policy deployment process, many other objectives are met, including improved communications throughout the organization, coordinated and aligned strategic goal-setting, developmental relationships between supervisors and subordinates, and a culture of continuous improvement.

Although it may seem that policy deployment should be implemented as a lean organization matures, in fact it is best initiated during the implementation process. Beginning to set structure and expectations, catchball, and roll-up processes during the change from mass to lean production sets the tone and creates new principles and processes for the new lean culture. The lean implementer would be wise to include policy deployment as one of his or her goals in the effort to become a lean organization.

Work Groups   10

By Ronald Link

People are the key to action in any manufacturing system. No matter how sophisticated or simple the situation, somewhere and somehow people are woven into the action equation. Whether it is pushing a button, sending a hand signal, activating a software program, or asking for help, any manufacturing process or system needs people. Given that fact, it is critical for people to offer their positive commitment to implement lean principles. Anything not supported eventually fails, and this is certainly true of the support required of people helping to move projects and systems forward. If lean implementers expect extraordinary results, they must find ways to activate extraordinary support. Average support helps yield average results, and poor support almost always ensures failure.

The most effective way to involve and engage people in the work force is through groups. The term "work groups" is used instead of "teams" because the groups discussed here are, in fact, brought together to work. Their work consists primarily of solving problems associated with their jobs. Their ideas are used to make life better at work and to continuously improve the way work is performed. They include virtually everyone associated with production, with the core of the work group made up of workers who add value.

Work groups exist to improve the amount of value-added work completed on a product. This is accomplished through the elimination of waste. Work groups are crucial to successfully implementing lean principles, and must be supported by everyone in the organization.

Volumes have been written about teams and the workplace. The author assumes that readers accept the premises that people are involved in all manufacturing systems and that positive involvement in those systems is good. This chapter discusses how to get a work force more fully engaged in the enterprise through work groups, which will help it move toward the goals of lean implementation and continuous improvement.

WASTE ELIMINATION

By now, it should be clear that lean manufacturing is really about the elimination of waste. Remember, there are seven basic forms of waste—overproduction, waiting, defects, movement/conveyance, inventory, overprocessing, and motion. Nowhere in the definitions of these wastes are people referenced. People are not waste. People can help control waste if they understand the parameters and rules of engagement. It is crucial when introducing the concept of lean manufacturing (in any organization and at any level) that the definitions of lean and waste be explored thoroughly. This should happen at the onset of discussions on why work groups are important to the implementation of lean principles.

Employees are concerned about job security. They want to know they will have a job tomorrow and the next day. When employees hear the words "lean" and "waste," they think the organization is using them to hide layoffs. If that idea is allowed to be formed and settle in, it will make the implementation of lean manufacturing much more difficult than it otherwise needs to be.

There are three basic reasons why people must not be included in the definitions of waste. First, as stated previously, people are not waste. Many mission/vision statements make people the company's top asset. This thought must be clearly conveyed to people at all levels of the organization. Second, not including people as waste is the right thing to do philosophically. In this society, people constantly strive for improvement. To think of others or ourselves as a form of waste contradicts our efforts to improve. We value life, and, therefore, we always strive to make things better. Why bother if people are waste? Lastly, there is a pragmatic reason for not thinking of people as waste in industry. People usually will not voluntarily erode their numbers. The first idea from a group resulting in treating people like waste in any way, such as laying them off as a result of a good idea, will likely be the last idea from the people involved. If managers want people to behave like they are adding value, they must treat them as if they are.

Waste is anything that adds cost but not value to a product. Sometimes it is helpful to think of it this way: if a customer could order a new product and follow it through the entire manufacturing process, which steps would he or she be willing to pay for? Which of the many steps would the customer view as not adding value (waste) to the final product? If there are too many people in the process, some of them not fully occupied, should they be considered waste? The answer is no. This major point is discussed in more detail later in this chapter.

Definitions of waste and lean do not preclude normal volume adjustments, attrition, and ongoing efficiencies. These are necessary to stay in business, just as being lean is necessary. In large plants and companies, sometimes it is difficult to determine the number of people associated with various projects and programs. Though it can be difficult, this determination must be made and then communicated thoroughly and openly. Organizations must be willing and able to share their plans for people, including reductions brought on by attrition, changes in production volumes, and other efficiency actions. Manpower savings generated by groups in the lean implementation must be clearly identified as still on the rolls, in an equal position with at least the same pay as before any reduction.

ASSUMPTIONS ABOUT PEOPLE

Before proceeding, there are four assumptions about people at work that underlie our comments about their involvement in a production system. These assumptions will sound a lot like the "soft side" of management. They are, in fact, on the soft side of the management ledger, but are also borne of first-hand knowledge and experience with people. If the following assumptions about people are correct (and there is an abundance of research and experience that states they are), then why isn't every organization performing at a high level? It seems like an easy thing to do. How can an organization tap into the thousands of ideas people have about making their work better and making life at work better? How can leadership in an organization capitalize on the power of people? The answers to these questions will be explored after a brief discussion on the four assumptions about people.

Assumption 1: People Want to be Treated Like Adults

People can handle both good and bad news about their jobs. They want to be told the truth and can spot lies and/or deception in an instant. Sometimes, managers think people cannot handle the truth about a given situation. In reality, they have probably handled much worse news in the course of their lives and can surely handle job-related bad news. People want to be heard and be given choices. It is difficult, maybe even impossible, to tell everyone everything and to give people choices about everything; nor is this approach advocated. However, managers and leaders should always operate under the assumption that the people in their work force are adults and want to be treated as such.

Employees want to be told the truth, they want to be heard, and they want choices. If an organization has a history of a parent-child relationship, a highly autocratic leadership style, or poor work force relations, some thought must go into the advent of this assumption. In fact, a good deal of work with management and/or union leaders might be necessary to accept this assumption.

Assumption 2: People Want to Do a Good Job

Generally, people understand their futures are somehow intertwined with their work. They want to do a good job to help ensure a bright future for themselves and their families. Only the rarest of people drive to work in the morning thinking "how can I mess things up at work today?" Most people never think that way. Instead, most people want to do a good job, ranging from extremely motivated people who want to be CEO some day, to others who simply want to do their assigned tasks as best they can. Also, people tend to take pride in their work. They want to feel needed and necessary.

Assumption 3: Those Close to a Problem Can Probably Solve It

It is common sense that people assigned to a specific job know that job better than anyone else. For many years that knowledge has been left untapped as engineers, managers, and supervisory personnel did the problem-solving. Fortunately, many companies are beginning to understand the problem-solving power of the people closest to a problem. People have ideas about how to improve their jobs as well as their life at work. It makes good business sense to engage those closest to a problem to try to arrive at a lasting solution. The fix then belongs to the people, and that ownership is priceless.

Assumption 4: People Will Help

If people know what the organization wants, they will help get it done. Business goals, competitive positioning, and wellness of the business are all important concepts that must be shared with the work force. General information, and how people can use it, must be communicated properly to help result in improvements.

To give a work force lots of good information and tell them they are empowered to act on it is not good enough. People must see value for themselves as individuals or as a group to bring the information to life in a positive way. Organizations obviously must not wait until they are in serious financial trouble to begin trying to communicate clearly with the work force. Employees will think it looks and feels like manipulation. Employees must be treated like adults and be given good and bad news honestly and consistently. A crucial factor in this asumption is that people do what they see their leaders doing, not necessarily what they are saying. In other words, walking the talk is the only way to prove to the work force that the words are real. Merely providing information to the work force is not enough. The work force must understand and see value in it to make the effort worthwhile.

WORK GROUPS PROVIDE THE BEST APPROACH

Work groups are the best way to optimize employees' talents and skills. There are many definitions of work groups, but this short and simple one meets the needs of lean manufacturing: A *work group* is any two or more people working together for a common cause. The term "work group" is used because it more closely connects with the manufacturing environment than the term "team," although the two terms can be synonymous. The definition of work groups includes other key characteristics, such as:

- they are long-term and well established;
- they are interested in improving their life at work through improvements on the job, and
- they want to improve the competitive positioning of their company by identifying and removing waste. (Remember, the definition of waste does not include people.)

Later in this chapter, the anatomy of a work group is examined more closely.

Pitfalls to Avoid

Attempts to overlay work groups on the work force usually yield minimal results or are outright failures. Research and on-the-job experience conclude that when work groups are properly installed, trained, and used, they improve individual, collective, and corporate performance through productivity improvements. Unfortunately, there are plenty of examples in which organizations claim to have work groups in place, but in fact they do not. Instead, they have work group shells in place. This unfortunate situation can happen for several reasons.

One reason may be that the company leader declares that the organization will have work groups. The employees responsible for such things dutifully go about arranging the groups. Very little, if any, training is done, and not much explanation may be offered as to why groups are being formed. This is not only a recipe for failure in the immediate effort, but it guarantees difficulty when arranging work groups the proper way in the future.

Another reason the "let's do work groups" syndrome occurs is because the organization is in trouble financially and/or competitively, and company leaders decide they must get the work force involved to survive. Sometimes this works for a while but can lead to problems. Employees may ask the leaders "where were you earlier?" or "why am I important now but you never had the time of day for me before?" These are good, fair questions from the staff when it is suddenly told it must form work groups for the company to survive.

The "let's do work groups" syndrome also can happen because it simply seems like the right thing to do. There are countless books, videos, and seminars available today extolling the virtues of work groups. Most of these efforts are well-intentioned, and many are quite good, but unless work groups are installed for the right reasons they will fail. Remember the assumption about people wanting to be treated like adults, and if they are, they will behave like adults. They will make adult decisions that positively influence their work, company, and future.

Work groups are created to solve problems and handle the increasing complexity of work. Using work groups to help resolve complex assignments is not new, but the power of groups is needed more than ever to solve today's increasingly complicated and demanding array of business and work problems. To believe that a small group of managers can most effectively cope with and solve all these problems is pure folly. Managers need all the help they can get.

WORK GROUPS AND THE WORKPLACE

Work groups can help individual employees optimize their gains through being an active, contributing member of a work group. Individual contributions on the shop floor sometimes get lost or minimized, but synergy created in work groups often makes individual efforts and gains even greater than they would have been on their own. Workplace culture, the force that dictates how to approach and use work groups—at least in the beginning of lean implementation—is based on the history of that workplace. In American manufacturing companies, the culture is often based on a "we-they" attitude. This can occur in both union and non-union shops. Often, management is seen as the adversary. The central reason for this view is lack of trust. Trust is the central issue in any change effort. Moving to a work-group-based organization is no exception.

Comfort Zones

Employees become accustomed to doing things a certain way, including how to do their work. They get in a rut—a comfortable rut—and it is difficult to change to a new pattern of behavior. Employees wonder if a change is really good for them. Management says it is, but so what?

Managers may have told the work force lots of things in the past that were short-lived. How do employees know the change will last? Will it be another "flavor of the month" concept? These questions involve moving from a comfort zone through the transition required for change. The leader refers to his or her "team," but employees may really think the leader means his or her "way." Leaders seem to refer to teams when they are in trouble, so employees know this too shall pass. It is difficult to move to a new way of working or a new pattern of behavior. Skepticism surrounds the reasons for change, and personnel wonder if the latest change will last more than a few months. Added to all this is the fact that some employees are a little nervous about the freedom that work groups can bring. The longer they operate in their comfort zone, the more difficult it is to move out of it. Despite all these issues and concerns, work groups are a powerful force to advance company and individual goals if they are installed, trained, and used properly.

ANATOMY OF A WORK GROUP

What does a good work group look like? What is its anatomy? The ideal size is an average of four to six workers, but it should have no fewer than two and no more than 15. The number of members in a given work group can be determined by the number and type of problems, work processes, geographic boundaries, or simply by stating how many people should be in an average work group. There is no magic formula for determining who lands in a specific work group, as there are often many variables involved.

The most important variable, however, is the number and type of problems. After all, the primary purpose of work groups is to solve problems. For example, launch areas have more problems than stable lines, so more work groups might be in order for launches.

Group Leaders

Work groups should have leaders. Leaders are often full-time, but can be part-time if necessary. This is best determined by team size, as well as the number and type of problems to be solved. Leaders may get extra pay for their responsibility, which must be determined early in the work group formation. If leaders are to receive extra pay for their efforts, their duties must be clearly spelled out (this must occur regardless of the payment issue). A solid, well-thought-out process must be used to identify and select leaders. The selection process can range from sophisticated assessment centers to a work group election.

Leaders should know all the jobs in the work group well so they can support people on those jobs and solve problems. This knowledge and ability requires time to acquire. Rotating a work group's leaders can seriously undermine this requirement. However, each organization must determine the best course for its unique situation. Whichever selection system is used, it must be instituted well and used consistently.

A de-selection system often may be necessary. Managers do the best they can when selecting leaders, but sometimes things do not work out, so there must be an agreed-to method for taking a leader off the job. Managers and work group members must answer questions in advance about what behavior calls for removal, whether or not to take pay away, etc. These are important issues, and must be dealt with early in the work group implementation process. The best approach is a positive one from the beginning, including training, coaching, and any counseling required along the way.

TRAINING

Work group training is a strategy aimed at overall organizational improvement. Effective training is grounded in the philosophy that by upgrading skills of the individuals in the company, the company as a whole is being upgraded as well. Training comes from three sources:

- Classroom training is used most often but probably is the least effective of the three types. Classroom training is good for establishing concepts and for reflecting upon other learning activities. It is usually the most expensive so it should be carefully planned. The learning environment is important in classroom training. Quality instructors are a necessity, yet transferring knowledge is quite difficult. Managers look for changes in behavior on the facility floor, and they must insist on the same in the classroom.
- Self-initiated learning activities are cheap and can be effective, but the company must rely on the individual to take the initiative for his or her learning. If this form of learning occurs in an area where the individual has a natural interest, it helps create an environment where successful learning can occur. The key for the company is to create the proper environment for successful learning. Companies effective at this training usually provide space and occasional financial support to ensure success.
- On-the-job training is the most powerful of the three forms of training. It allows the real situation to create, and in a substantial part govern, the learning. On-the-job training is the most relevant to the individual, and is usually the most effective at providing substantive learning. On-the-job training does not have to be ad-hoc, and it can be highly structured to achieve the desired learning objectives. It can range from something as simple as a self-guided tour to a highly sophisticated program. It is common for individuals in the company to claim that their most powerful learning experiences have been on the job.

Proper Training is Essential

Work groups, like individuals, must be well-trained. Some organizations have made a critical error by not training people to accept new roles and responsibilities. In the early 1980s, many companies attempted to involve the hourly work force in the shop floor management of the business. Some companies, and some plants within companies, had success with this effort while others did not. There are many reasons for the success and failure of this effort. One way to practically ensure a failed effort was not training the work force in ways to accept its newly appointed roles and responsibilities. Telling a group of people they are a work group and that they will have some say in how their department and the company operate, and then not providing at least some training, is unfair to all parties and probably will cause the effort to end in failure.

Necessary Skills

Work group members need two types of skills—technical and interpersonal ("soft" skills). Training will vary for each organization based on training history, status of lean implementation, etc. It is estimated that a minimum of 54 hours per employee is required for technical training, as outlined below:

- total productive maintenance—12 hours,
- visual factory—8 hours,
- quick changeover—8 hours,
- mistake proofing—8 hours,
- standardized work sheets—8 hours,
- measurables—8 hours, and
- total cost management—2 hours.

Each of these subjects (except perhaps total cost management) should be taught in more than one session. The estimated hours do not include time for course development or refresher training. In addition, there may be other courses/skills in which an organization wants to make sure its employees are trained. It is not unusual for an organization to present 70-80 hours of technical training per employee to help get lean implementation moving forward and building a foundation for sustaining it.

Soft-skills training is also crucial. Again, a thorough analysis is required to determine exactly what group dynamics skills are needed. The best estimate in this area, based on core skill requirements, is 38 hours of training per employee, which includes 16 hours of problem-solving training. Group dynamics or soft skills training includes:

- explanation of work groups—1 hour
- problem-solving—16 hours
- effective meetings—4 hours
- financial information—1 hour, and
- group dynamics—16 hours, which may include four hours each on conflict management, decision-making, roles and responsibilities, and personality differences.

Again, these estimates are based on experience, but any organization's needs can vary. Careful analysis is necessary.

WHO TRAINS AND WHO GETS TRAINED

Decisions on who should be trained and who should conduct the training have considerable cost and time implications, but they also have significant employee relations implications. Work group members should receive, at a minimum, all of the training in the previous section. This means their supervisors should receive the same training either before or during the work group's training.

Engineers must be considered for certain portions of the training so they are aware of what is happening on the shop floor. When engineers are aware of what is taking place on the shop floor, they are in a better position to help without feeling threatened. All of these needs, including what training managers should receive, will be identified through a good training analysis. This analysis also will help determine proper timing and budgeting. A solid training plan is imperative because time and cost constraints, as well as good training principles, generally do not allow all of the training to be done in one calendar year. Training should be done "just-in-time" when possible. Table 10-1 is a suggested training matrix for a manufacturing plant.

Organizations have two basic choices concerning who conducts training. They can use external trainers (for example, professional trainers and consultants) or internal trainers. There are several advantages to using internal trainers. They are:

- Cost—internal trainers are, in the long run, more cost-effective than outside trainers.
- Ownership—internal trainers must live with what they train.

Table 10-1. Suggested training matrix for a manufacturing plant

Skills	Work Group Members (hours)	Supervisors (hours)	Middle Managers (hours)	Union Reps (hours)	Top Managers (hours)
Technical					
TPM	12	12	12	12	12
Visual factory	8	8	4	4	1
Quick changeover	8	8	4	4	1
Mistake proofing	8	8	4	4	1
Standardized work sheets	8	8	8	8	4
Measurables	8	8	4	4	1
Total cost management	2	2	1	1	1
Group Dynamics					
Explanation of work groups	1	1	1	1	1
Problem-solving	16	16	4	4	4
Effective meetings	4	4	2	2	2
Financial information	1	1	1	1	—
Group dynamics	16	16	16	16	8

- Flexibility—using internal trainers greatly increases the options for when and how training can be accomplished, and can support the just-in-time concept.
- Knowledge of product—internal trainers know the company's product and processes, and can relate to questions/problems raised in sessions.
- Knowledge of organization—internal trainers have a better understanding of how the organization fits together.
- Jargon—internal trainers are familiar with acronyms and the special language that each organization develops over time.
- Personal growth—there are people with great training skills in almost every organization, and using internal trainers often gives these people a chance to blossom.

It may be wise to use external trainers to get started, but they should always train internal people to replace them. Other training considerations include budgeting, having a time-phased plan, arranging training so it is just-in-time if possible, and the need for refresher training. All of these factors can be planned fairly easily if started early in the process; each can cause considerable trouble if they must be reacted to without prior planning. Training can be expensive no matter who conducts and attends it, and it always takes time. It is, however, necessary. Therefore, good planning is mandatory.

WORK GROUP MEETINGS

Work groups must have regular meetings if they are to be effective. "Regular" can mean anything from daily to monthly. Weekly meetings are best unless the work group and its support system are mature. The meetings should be at least an hour in length because anything less is ineffective time-wise. If a half-hour meeting is scheduled, by the time employees arrive and get settled, that time is gone. This does not include daily start-up meetings held in many places. Start-up meetings often take no more than 15 minutes and are different from the meetings discussed here. Efficient meeting facilities must be available, and work group members may need training on how to conduct effective meetings.

Meetings should be mandatory—voluntary meetings invite trouble. Work group meetings should be required during the workday. If a meeting is not important enough to require attendance, then the need for the meeting should be seriously questioned. Of course, if meeting attendance is required, the time must be paid. Sometimes overtime is necessary, but usually an organization can find a way to have meetings on straight time. Mandatory attendance is different from mandatory participation. People usually cannot be made to participate. The best strategy to deal with someone who refuses to participate is to talk about his or her job. That usually brings some participation from even the most reluctant work group member.

Another important matter to decide is whether work groups should be semi-autonomous or self-directed. Not many organizations are ready for truly self-directed work groups. Semi-autonomous work groups (those with parameters and some organizational oversight) present many concerns and issues without planning for them to be totally self-directed, especially at first. An organization thinking of self-directed teams should examine its plans closely.

WHERE TO BEGIN?

With all the questions and concerns, as well as positive aspects of work groups, how does a manager begin implementing them? The first step is to assess the leadership's readiness for work groups and/or a work-group-based organization. This can be done through simple paper and pencil surveys, such as those in Figures 10-1 and 10-2. Results from the survey in Figure 10-1 can be used by the leadership group to determine where it stands on the question of work groups. Results from the employee readiness survey (Figure 10-2) will help leaders determine if their departments or areas are ready to implement work groups.

Answers to questions such as "why do we need work groups?" and "how will having work groups impact our organization?" can be identified as a result of the survey and ensuing discussions. Discussing opinions about these questions and others before implementing work groups is crucial to their success.

Leadership includes key managers and union leaders involved in lean implementation so the survey should be completed jointly. This is an excellent way to open the lines of communication about a subject that often has emotion attached to it. Whether or not a union is present, a widely trusted person in the organization should tabulate the results and report on them. The information should be kept within the work group until it decides how and when to release it.

Next, the leadership group should begin working on the results of its survey (Figure 10-1). The survey helps answer the basic question of whether the leadership group wants and is ready for work groups. The survey should also reveal items to be addressed,

Objectives	Candor	Instructions
To determine enablers and barriers to the implementation of work groups. To provide information for development of an action plan for the implementation of work groups.	As with any survey, information resulting from it is useful only if participants are candid with their responses. To not be candid yields information that not only is waste but can be harmful. The survey should be completed anonymously.	Please circle the number that best fits your opinion on each of the numbered items: 4 = Strongly agree 3 = Somewhat agree 2 = Somewhat disagree 1 = Strongly disagree

Item	Rating
1. Management in this organization will provide consistent, sustained support for implementation and effective use of work groups.	4 3 2 1
2. This organization has a history of successfully making major changes.	4 3 2 1
3. Most supervisors and managers in this organization have the knowledge and skills needed to implement work groups.	4 3 2 1
4. Management will make business decisions (training, budget, time for meetings, etc.) to ensure the success of work groups.	4 3 2 1
5. This organization has the resources required to support the implementation of work groups.	4 3 2 1
6. Work groups are important to achieving our business objectives.	4 3 2 1
7. Management will devote a substantial amount of its time to the implementation of work groups.	4 3 2 1

Figure 10-1. Management readiness survey for work groups.

Item	Rating			
8. I have the personal/professional knowledge and skills necessary to support work groups.	4	3	2	1
9. My boss is committed to work group implementation.	4	3	2	1
10. The management style in this organization will effectively support work groups.	4	3	2	1
11. Most supervisors and managers in this organization understand what their job responsibilities will be when work groups are implemented.	4	3	2	1
12. Work group implementation will require me to change my behavior.	4	3	2	1
13. First-line supervisors will support work groups.	4	3	2	1
14. I am clear on what my job responsibilities will be when work groups are implemented.	4	3	2	1
15. Management's relationship with the general work force will facilitate the implementation of work groups.	4	3	2	1
16. Managers in this organization trust each other.	4	3	2	1
17. Management provides accurate, useful information to the general work force.	4	3	2	1

Question	Area
1	Management commitment
2	Change history
3	Skills
4	Management commitment
5	Resources
6	Business case for work groups

Question	Area
7	Management commitment
8	Skills
9	Management commitment
10	Behaviors
11	Roles and responsibilities
12	Behaviors

Question	Area
13	Behaviors
14	Roles and responsibilities
15	Behaviors
16	Trust
17	Management commitment

Figure 10-1. (continued) *The last section of the survey should not be distributed with the survey. It is intended for use by a facilitator, including discussion of which questions deal with the various areas identified.

Lean Manufacturing: A Plant Floor Guide

Objectives	Candor	Instructions
To determine the readiness of a given department or area to implement work groups. To facilitate the development of an action plan.	As with any survey, information resulting from it is useful only if participants are candid with their responses. To not be candid yields information that not only is waste but can be harmful. The survey should be completed anonymously.	Please circle the number that best fits your opinion on each of the items: 4 = Strongly agree 3 = Somewhat agree 2 = Somewhat disagree 1 = Strongly disagree

Item	Rating
1. I want more participation in decisions affecting work.	4 3 2 1
2. The overall climate in my area is good.	4 3 2 1
3. We produce good quality parts in my area.	4 3 2 1
4. I trust information I receive from management.	4 3 2 1
5. I feel I can talk to my supervisor about any part of my job.	4 3 2 1
6. Ideas for making things better are shared in my area.	4 3 2 1
7. I believe we should implement work groups in my area.	4 3 2 1
8. Management trusts the people in my area.	4 3 2 1
9. I know the goals for my area (quality, safety, production, etc.).	4 3 2 1

Figure 10-2. Employee readiness survey for work groups.

Item	Rating
10. I receive constructive feedback from my supervisor.	4 3 2 1
11. Adequate information is made available to the people in my area.	4 3 2 1
12. My supervisor encourages participation in decision making.	4 3 2 1
13. People in my area understand information made available to them.	4 3 2 1
14. My ideas are considered.	4 3 2 1
15. The people in my area trust management.	4 3 2 1
16. I have all the tools I need to do my job.	4 3 2 1
17. I trust my fellow employees.	4 3 2 1
18. People in my area use information made available to them.	4 3 2 1
19. My area uses a consistent problem-solving process.	4 3 2 1
20. I think my work area has good productivity.	4 3 2 1

Figure 10-2. (continued)

regardless of the status of work groups. If leaders determine they want to initiate or revitalize work groups, they should develop a business case for doing it.

Lean manufacturing work groups do not exist so people can feel warm and fuzzy. They exist for the betterment of the business, including the employee's life at work. If business reasons for work groups are satisfied, and work group members feel warm and fuzzy along the way, great. Reaching business goals is the objective, not feeling warm and fuzzy.

A steering committee must be formed at this point in the implementation process. Next, the staff should be told of the decision to establish work groups. Often, organizations do not have all the details worked out at this point, but that is okay. Initially, employees should be informed of the work groups through a series of meetings, not through the company newsletter—that comes later. In these meetings, people should receive the following information:

- what the business case is for establishing work groups;
- no layoffs will occur as a result of establishing work groups;
- an explanation of the pilot group and how it was selected, and
- an examination of why work groups are the best way to identify and eliminate waste.

Union leadership (if applicable) should be present at the meetings, and, if appropriate, participate in the presentation.

Leadership must be prepared to begin work on issues uncovered in the survey and those brought up in the work group announcement meetings.

STEERING COMMITTEES

A steering committee is crucial to developing effective work groups. This committee must be a joint effort if the shop is union, with an equal number of representatives from union and management. The committee should be co-chaired by the two respective leaders. This committee must meet regularly, probably weekly at first, and should publish notes from the meetings and make them available for everyone in the organization to read. The steering committee may need some training on how to conduct effective meetings. It should also develop a work force communications plan. Most importantly, the steering committee must be willing to take on difficult issues and resolve them. The following is a list of typical steering committee functions when dealing with the implementation of work groups:

- Oversee the implementation of work groups.
- Ensure proper use of resources.
- Define work groups and their boundaries.
- Develop a feedback system for performance.
- Determine use of work group leaders, if applicable.
- Define roles and responsibilities for all stakeholders.
- Oversee the implementation committee.
- Manage the growth of work groups.

There may be additional duties at each local level. The steering committee should be sized properly—large enough to have the right representation, but not too big to function effectively.

PILOT AREAS

Work group implementation should begin slowly. The best way to begin is by using a pilot area or department. The work force in the pilot area must be willing to accept its role as a pilot for work groups. Employee readiness can be checked through the use of a simple survey (Figure 10-2). Using survey results and the knowledge of its work force, a leadership group can determine whether the pilot area is ready to be used as an example of lean implementation for the rest of the organization. It is critical to be prepared when delivering training to the pilot area. It is equally critical to have support mechanisms in place, such as a follow-up system, to handle concerns raised by the pilot group. Supervisors must be included in this entire process. They may feel threatened as it is, and to not include them in the planning and implementation actions will probably alienate them from the work group. They know a great deal about life on the shop floor, and they possess a wealth of information about how to make things work.

IMPLEMENTING WORK GROUPS

The best place to begin a work group implementation is in a designated pilot area. This area provides a place to test ideas, discuss the process with participants, and learn from trying. This should occur during the stability phase of the five-phase implementation model (discussed in detail in Chapter 7). It is not necessary to have formal work groups (work groups with formal rules or guidelines) at the outset. Sometimes it is better to let them evolve at their own pace. Organizations should progress only as fast as the people involved want to accept.

Some parts of the implementation process can begin to go plant-wide in Phase II (the continuous flow phase of the overall implementation). At this point, formal work groups begin to work with each other, as well as some cross-functional groups. The need for smooth-functioning work groups accelerates as the lean implementation progresses, so a well-thought-out, continuously reviewed plan must be developed.

Issues Faced by Work Groups

Now that the basics are in place and the organization is ready to launch work groups, project managers should review issues the groups will confront as they begin to implement lean concepts. Nine of the most common issues are outlined below:

1. Unstated, unclear or misunderstood goals—goals for the work group must be clear. The steering committee must set the organization's overall goals regarding work groups, but the pilot work group is responsible for setting its own specific goals. The goals should be written and be somewhat difficult to achieve, but not too ambitious.
2. Trust is always a concern. The only way to earn trust is to start with honesty. Honesty brings integrity, which brings trust.
3. Communications—organizations, and especially managers, often have difficulty letting go of information. Organizations should treat their people like adults and let them know what is going on in the company. People often wait for someone

else to do the communicating. In any case, management should err on the side of over-communicating. Communications should add value.
4. Unclear roles—clear roles are a necessity. The steering committee, supervisors, and work group members, along with any other key stakeholders in the organization, must have clear, written roles and responsibilities.
5. Work group leaders—if managers want their work groups to have leaders, should they get paid extra? Should they be full or part time, elected or appointed? How long should they be in office and do they need special training? How is a leader who is not working out removed? All of these questions must be decided before the work group begins. This is another job for the steering committee.
6. Problem-solving is the hub around which the work groups in a lean organization revolve. A consistent problem-solving process should be used during work group meetings. The process should be easy to use, and some training may be required. The work group should solve problems, and a disciplined approach toward the problem-solving system must be followed.
7. Different strokes—there will be different personalities in the work group, and each is important. The work group should learn the different personalities present and know the value each adds. Remember, people are not waste.
8. Decision-making—a fully understood operational decision-making process must be in place, and the process must be used consistently and with input from everyone. This area can be frustrating because many people or work groups have a strong tendency to jump to action with just a little information. Remember, it seems difficult to find the time to do things right the first time, but we always seem to find time to do them over.
9. Vision—does the work group know where it is headed? Is its destination the same as the organization's? Vision provides something to hold onto when the going gets tough.

Work groups are not alone when having to face issues. Organizations implementing work groups have issues as well. Here are some of the issues faced by managers, supervisors, union officials, and others charged with overseeing the implementation of lean:

- It is critical for work groups to eventually control the design of their work. This is often difficult for engineers to accept and can cause conflict.
- The current compensation plan may be challenged as work groups are established. The steering committee must make decisions in this arena early and clearly communicate them.
- Supervisors might be threatened by the advent of work groups. They might be confused about their role, and even their job security. They must be given the same commitment as the hourly work groups—no layoffs as a result of work group implementation. This issue must be resolved early in the implementation process.
- Sometimes skilled tradespeople want to be a part of work groups, and sometimes they do not, but they must be involved in some way. Three possibilities include having them as an integral part of the work group, being outside the work group itself but dedicated to serving it, or as a member of a skilled trades group.
- If there is not enough support for work groups, do not implement them. Support means giving up some power, spending money, listening, holding meetings, sending e-mails, and visiting the work groups. Engagement means being interconnected with the work groups, understanding one's role and carrying it out, and knowing when and when not to intervene.

GREENFIELD VERSUS BROWNFIELD

The methods for implementing work groups can vary significantly between greenfield and brownfield sites. People involved in establishing an organization from scratch (a *greenfield site*) have a wonderful opportunity to create a work-group-based organizational culture if they choose, and they should. Once leaders decide to create a work-group-based organization, they can use a selection system that supports the kind of organization they want. They have the opportunity to clearly delineate roles and responsibilities as people are hired and trained. If done properly, establishing a greenfield site should result in a young culture that integrates work groups as a preferred way of life at work.

Greenfield site managers should have a work group implementation steering committee to deal with important questions, such as roles and responsibilities for work group members and leaders, compensation matters, and a problem resolution process. Implementing a work-group-based organization at a greenfield site requires considerable up-front planning, but should result in a smooth running operation. Also, constant care and feeding of the work groups is required to sustain them in a value-added mode over time.

Establishing and implementing work groups in a *brownfield site* can be an entirely different matter. Brownfield sites (sites with a history and established culture) can be difficult to change. This is particularly true if the organization has experienced one or more false starts with work groups, has a trust issue, or is doing well financially to the extent that people see no reason to change.

The question of "what's in it for me?" is prevalent in brownfield work group implementation, and change agents must be able to answer it sufficiently to have a chance at implementing lasting work groups. Actions such as ensuring commitment to change, establishing a steering committee, making decisions regarding leaders, etc., are critical.

While greenfield and brownfield sites bring their own special sets of problems and issues, both are usually good candidates for work group implementation. The most important aspect of greenfield implementation is deciding what kind of culture the organization should have and (assuming work groups are included) going about thoughtful implementation from the beginning. In brownfield sites, the most important aspect is deciding to change and sustaining that decision through continual efforts (including the implementation of work groups).

CONCLUSION

Even with all the issues involved in implementing and maintaining work groups, they are well worth the effort. They produce better results in the critical areas of quality, cost, productivity, delivery, and employee satisfaction. They make work better, and they make life at work better. Lean manufacturing implementation must include the minds as well as the hands of everyone. The best way to engage everyone is through the use of work groups. How else will the implementation generate the thousands of ideas required for continuous improvement? If managers want everyone's help, and realize the best way to get it is through work groups, then they must be willing to fully support them. Without support, any structure will fail. With support, work groups will definitely help the organization to succeed in this uncompromisingly competitive world.

Visual Factory 11

By Gregory Thomerson

A *visual factory* is a system of aids used to organize and control the workplace environment, ensure consistent quality, and provide support for productivity standards. It promotes effective communication throughout the organization by creating visual factory language for the workplace. Creating an effective visual factory is vital for communicating the current situation within a lean production process. Visual language enables operators and managers to quickly distinguish between the desired situation (the standard) and abnormalities in the manufacturing process.

Once a visual factory system is in place and all parties are accustomed to its elements, it is easy to notice when current production is operating abnormally. Anyone is able to identify when waste is present and knows how to return to the standard operating procedure. A visual factory enhances the ability of operators, managers, supervisors, and visitors to grasp the current operational condition.

In a visual factory workplace, every item deemed necessary has a designated location and remains there except when in use. This promotes orderly self-cleaning conditions on the production line. Work activities are standardized and are clearly communicated to everyone. Work is organized at all times, not just periodically in an effort to prepare for a review.

BENEFITS OF VISUAL FACTORY

A visual factory system has many benefits. It:
- promotes zero defects—conditions for zero defects, zero failures, and zero waste are established and maintained by all.

- shares information—it makes visible all information needed for effective control of storage, operations, equipment quality and safety, current status, and improvement targets.
- alerts everyone to abnormalities—every standard must be easily visible so that any deviation or abnormality can be instantly detected.
- aids in quick recovery—because all information and standards are so easily visible, abnormal conditions are corrected quickly.
- promotes prevention—it helps deter abnormalities before they occur, rather than constantly correcting them.
- eliminates waste—waste that consumes effort and resources but adds no value to the end product is exposed.
- promotes worker autonomy—adherence to standards increases when everyone sees how the system is intended to function.
- supports continuous improvement—stakeholders can look closely at each part of the process and find more opportunities for improvement.

VISUAL FACTORY AND THE ELIMINATION OF WASTE

A primary goal of lean implementation is eliminating waste. A visual factory system is helpful in identifying waste (defects, waiting, motion, overprocessing, overproduction, inventory, and inefficiency) and pointing the way to eliminating it.

The following are some real-world examples of waste eliminated by implementing a visual factory system in various facilities:

- In-use floor space was reduced by 42%. By reducing the amount of utilized floor space, the waste of motion space was reduced as well.
- Twelve Occupational Safety and Health Administration (OSHA) violations were removed in one cell. Safety was improved and people worked more efficiently.
- Non-machine equipment was reduced by 45% in a 4,000-ft^2 disassembly area.
- Total operator handling time of inventory was reduced by 20 hours a week in a department of five employees. This reduced wasted motion as well.

- Subassembly procedures were standardized and posted over three shifts. Standardizing subassembly procedures lessened defects, waiting, motion, and inefficiency because procedures were performed faster and better.
- 1,000 ft^2 of storage space was removed. Storage space was reduced because inventory was reduced.

HOW TO USE A VISUAL FACTORY SYSTEM

There are three main elements in a visual factory system, as shown in Figure 11-1.

1. The foundation is workplace organization and standardization.
2. Information sharing with visual displays.
3. Preventing abnormalities of all kinds through visual controls and error-proofing devices.

As these elements are discussed, prospective lean implementers will see how each can be broken down into finer components.

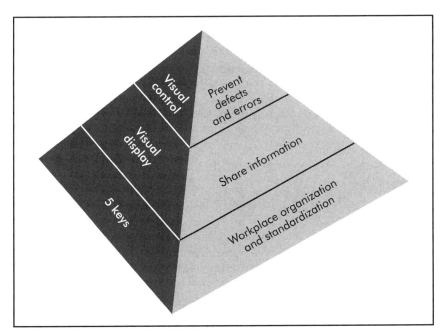

Figure 11-1. The visual factory pyramid.

Workplace Organization

The foundation of a visual factory is workplace organization. Organizing the workplace leads to stability in the production process. Establishing a visual factory during workplace organization involves working through the following five keys, also referred to as the 5-S strategy:

- Sort (organization)—nothing is extra or unnecessary.
- Stabilize (orderliness)—a place for everything and everything in its place.
- Shine (cleanliness)—the environment is immaculate and self-cleaning.
- Standardize (adherence)—standards are easy to see and understand.
- Sustain (self-discipline)—pride is created in the workplace, as well as the discipline to maintain it and to continuously improve.

Sort (Organization)

Sort through items in an area. Keep what is needed and eliminate what is not. Reduce the number of items to those actually needed. Find alternate storage for tools, parts, equipment, and supplies that are needed but not used daily. Find an appropriate location for all items according to their usage, urgency, frequency, size, and weight. Determine how to prevent the further accumulation of unnecessary items. This practice is called red tagging.

Red tagging visually identifies what is not needed in the workplace. Red tags ask, "why am I here?" Begin red tagging by first establishing rules for what is needed and where it belongs. Attach red tags to all unnecessary and out-of-place items. Write the specific reason for red tagging, then sign and date each tag. Remove and store red-tagged items in a temporary holding area.

A *holding area* is temporary storage for items that must be removed from the work area but cannot be discarded until interested parties have agreed upon the disposal. A local red-tag holding area should be set up in each department that does red tagging. A central red-tag holding area may be established for the entire plant, for items whose final disposition cannot be decided or implemented at the local work area.

Holding areas (local and central) should have designated managers who log and monitor red-tag items entering and leaving the holding area. They should be cleared in a timely manner—daily, weekly, or monthly—according to an agreed-upon standard. Sort through and dispose of items that are truly unnecessary. Prepare all other items for relocation. Make sure that all interested parties agree. Find ways to improve the workplace so unnecessary items do not accumulate. Red tagging should be practiced and continued on a regular basis.

Stabilize (Orderliness)

Stabilizing the process and process area ensures that there is a place for everything and that everything is in its place. Determine the best location for all necessary items to be kept in the area, and how many of each will be stored. Then, set limits on the space allocated. This increases job efficiency by making it easy for anyone to find, use, and return items.

The first step in stabilizing the work environment is to document the current situation by creating a location map, as shown in Figure 11-2. The purpose of a location map is to determine where equipment and supplies in the work area are and to show the flow of activities between these areas.

Develop location indicators as illustrated in Figure 11-3, and establish appropriate locations for all needed items using location

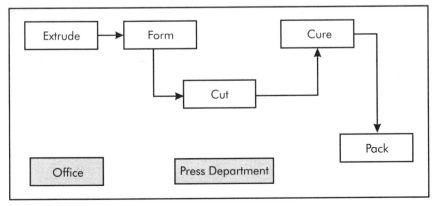

Figure 11-2. Location map.

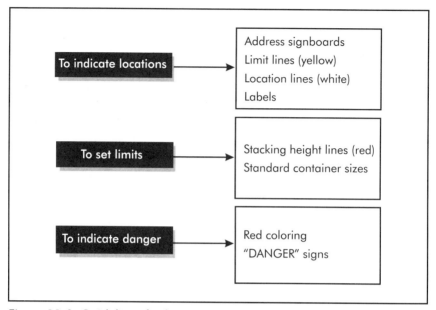

Figure 11-3. Guidelines for location indicators.

maps with process flow arrows. Prepare locations or containers for needed items, but limit space physically and visually to prevent accumulation and use containers that only hold the standard number of items. Make and post addresses for each item. Addresses should include the correct location, title, and quantity. The correct location is where the item is kept. It is typical to use a column, rack, or bin number. The numbering system must be user-friendly and something with which all users are familiar. The correct title is the appropriate name and part number of the item. The correct quantity describes a minimum and a maximum number for on-hand inventory. The minimum indicates the number on hand that signals to the users and material delivery personnel that more parts/items need to be brought in. This is done by putting in a Kanban (for a full discussion on Kanban, see Chapter 17) to order more of the same item. The maximum number is simply the amount of units that should be in the location when it is full. Try to keep approximately two hours' worth of the item/part in stock if the material delivery schedule can support a two-hour delivery of parts.

Location indicators should continually be evaluated and improved. To indicate locations, use address signboards and place them directly above the part to show its correct place. Limits are predetermined and should be marked to indicate the maximum level of the part on hand. Red height lines are used to indicate the maximum height when stacking items/parts. Sometimes this is accomplished by simply hanging a red bar held by a string on each end. Standard container sizes should be determined so that the part or item is stored only in the appropriate container. Danger points, such as pinch points or hazards, are designated in a manner so they are easily recognized and alert individuals to use caution.

Shine (Cleanliness)

Shine is eliminating dirt, dust, fluids, and other debris to make the work area clean. Every individual in the production process should find ways to keep the workplace clean at all times. For example, strategies should be developed to eliminate sources of contamination, such as controlling coolant overspray and containing cutting shavings as they are produced. Cleaning should be a part of everyday work for all employees. It builds pride in the workplace.

Shine is an important part of the preventive maintenance process. For example, cleaning and eliminating sources of contamination can be used as a method of inspecting equipment. Once the equipment is thoroughly cleaned, frequent inspections can spot potential problems before they result in production downtime.

Standardize (Adherence)

The goal of the fourth key is to achieve a state in which the first three keys are thoroughly maintained by everyone in the facility. Begin by establishing a monitoring and review procedure and educate everyone on its use. Consistently reinforce that the expected standard situation is to be applied in all work areas. Consistency is important if managers expect to be taken seriously. Standardize how the organizing and cleaning is done, and make standards visible so that any abnormal condition can be easily and immediately recognized.

Devise and maintain methods for adhering to the desired state and prevent deviations from the standard. Prevent the accumulation of unnecessary or unneeded materials in the work area. Ensure that everything is returned to its home. If everything has a place and everything is in its place, then any out-of-place item is recognized and the proper action taken to reinforce this policy/procedure. Maintain cleanliness by having all operators clean up after themselves. By taking ownership in the process, the desired level of cleanliness is maintained.

Sustain (Self-discipline)

When the sustain key is in place, correct procedures have become habitual. Workers are properly trained and everyone has adopted the visual factory system. The workplace operates in an orderly fashion according to agreed-upon procedures, and the plant's policies and procedures support the first four keys of the visual factory.

VISUAL DISPLAYS/VISUAL CONTROLS

Visual displays and *controls* are a system of devices, information, color-coding, layouts, and signboards standardized to create a common visual language in the workplace. They promptly distinguish between normal and abnormal conditions, make abnormalities and waste obvious enough for anyone to recognize, and constantly expose areas in need of improvement. Visual control falls into six natural categories, or stages. Each stage is progressively more powerful in its capacity to share information, alert everyone to abnormalities, speed recovery, strengthen adherence, and promote prevention. Heading upward on the visual factory pyramid shown in Figure 11-4, there is an increasing degree of visual control. Looking up the pyramid, visual displays are encountered first, followed by visual controls.

Visual displays communicate important information but do not necessarily control what people or machines do. For example, information may be posted on a safety chart that influences behavior but does not control it. Visual displays fall into the first two levels of the visual control pyramid.

Visual Factory

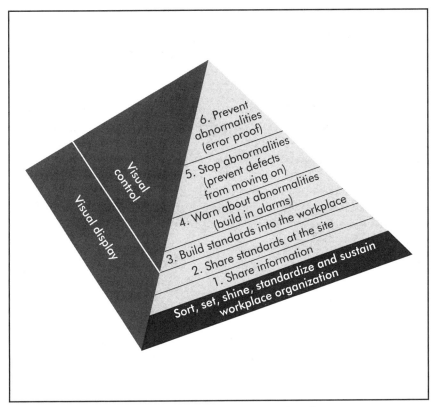

Figure 11-4. The visual factory pyramid in more detail.

Visual controls communicate information in such a way that activities are performed according to standards. For example, painted walkways, machine guards, and color-coded lubrication inlets all direct and control specific human behaviors. Visual controls fall into the top four levels of the visual control pyramid.

It is important to understand the differences between visual displays and controls so the most effective tool can be chosen to meet a specific plant-floor situation. In some cases, simply sharing information using visual displays can solve a problem. In other situations, a visual control may be needed to maintain on-site control of work activities.

There are six levels of visual display and control. Levels 1 and 2 pertain to visual display, and levels 3 through 6 cover visual controls.

Level 1

Level 1 visual displays involve sharing information about all activities, especially the results of control activities. This lets everyone see how closely performance conforms to expectations. Sharing information while it is fresh lets workers respond to trends in performance before negative patterns take root. One example is to post graphs indicating production performance in the work area.

Level 2

Level 2 visual displays involve sharing standards at the site. Sharing information about standard specifications and methods where it is needed lets everyone begin to identify nonconformance as it occurs and helps to correct it. For example, post instructions, procedures, and diagrams near the processes where they relate.

Level 3

A Level 3 visual control involves building standards into the workplace. An example here would be a red light next to instructions so that a driver can easily see both, like a status board with lights.

Level 4

A Level 4 visual control warns of abnormalities with something like a bell that immediately sounds when a car door is opened if the lights are on. This is like a status board with chimes and lights.

Level 5

Level 5 visual controls stop abnormalities with a device that prevents keys from being removed from the ignition until the lights are switched off. This is like a machine shut-down.

Level 6

A Level 6 visual control prevents abnormalities with something like a device that automatically switches off lights when a car engine is turned off.

Examples

Figures 11-5, 11-6, and 11-7 illustrate examples of visual displays and visual controls. The following are definitions of the different types of visual displays and visual controls.

- *Signboards* indicate what belongs where and in what amount, so anyone can understand where everything is supposed to be.
- *Material flow cards* are administrative tools that help manage just-in-time production. The two main types of material flow cards are transport and production cards. For more information on cards, see Chapter 17.
- *Status boards* immediately alert work groups and supervisors to abnormalities (danger) occurring on the factory floor, indicate where the trouble is, and show the line status at all times.
- *Red tagging* helps distinguish between needed and unneeded items in the workplace. Red-tag teams use red tags to mark unnecessary items for removal.
- *White-line demarcations* mark out pathways and in-process storage sites with white tape or paint making it easy for anyone to keep the work area clear and neat.
- *Red-line demarcations* on poles or walls next to stacks of inventory (warehouse or in-process) mark the maximum allowable stack height and easily reveal when excess inventory has accumulated.
- *Standardized worksheets* are used to indicate the agreed-upon way to perform operations within the work area. These sheets should be on display in every work area.
- *Defective item displays* are set up in work areas where defects have occurred. They should exhibit defective items along with other data to aid in recognition and prevention.
- *Error prevention boards* document the types of errors that occur in the work area and highlight ongoing efforts to reduce them.

Setting Up Visual Displays

Visual displays are any type of device, aid, or sign that helps inform, direct, warn, or track. Setting up visual factory displays

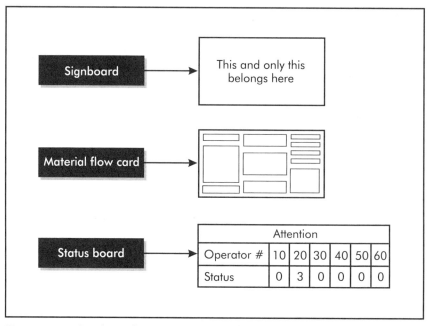

Figure 11-5. Signboard, material flow card, and status board.

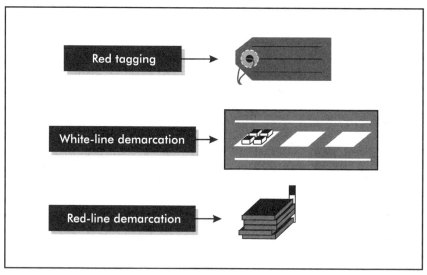

Figure 11-6. Red tagging, white-line demarcation, and red-line demarcation.

Visual Factory

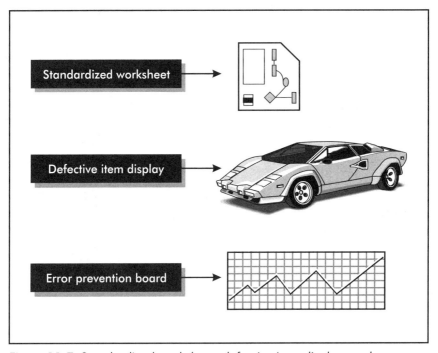

Figure 11-7. Standardized worksheet, defective item display, and error prevention board.

requires that standards be established before any other part of the process. Visual displays can then be used to show the goals and results of activities and improvements in productivity, setup time, defects, downtime, and throughput. They can include administrative and process charts, location indicators, boundary samples, maintenance and training schedules, and process jobs.

Managers must do more than post signs and tell workers what to do. It must be visually easy for people to know what comes next. To make visual displays effective, integrate standards into the workplace (see Figure 11-8 for an example).

Visual displays communicate vital information, but visual controls enable anyone to recognize important information, problems, waste, or deviations from the expected standard. Although the term "visual control" is used, such methods often relate to other senses, such as hearing, touching, smelling, and tasting.

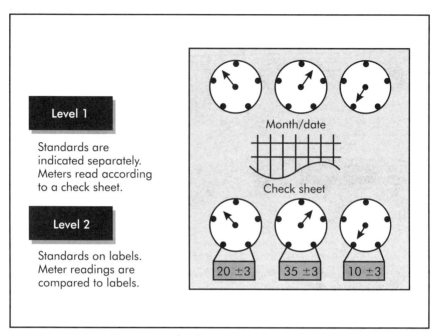

Figure 11-8. Integrating standards with visual displays.

There are many types of visual displays. In addition to the displays and controls described earlier, a few more examples include:

- *Storyboards* (wall displays)—teams can use a board with a combination of visuals, text, and charts to tell the story of an improvement process or project. Storyboards can be used to educate and inform, or to gather ideas and feedback.
- *Maps*—two types of maps are generally used. Location maps show the layout of all or part of a physical area, and process maps track activities and information about a process, procedure, or work flow.
- *Photos* are an excellent tool for documenting and communicating correct procedures and locations, or for demonstrating before and after information about improvement activities.
- *Checklists* are structured forms that make it easy to record and analyze data where it is collected in every stage of process improvement. The best checklists are easy to use and display data in formats that reveal underlying patterns.

- *Material flow cards* trigger part or supply re-ordering on a just-in-time basis. The material flow card communicates how many parts or supplies of a particular type are needed and where and when they are needed.
- *Improvement data and progress charts* posted in the workplace keep everyone continually informed about the progress of improvement activities.

Visual displays can solve many of the information and communication problems that prohibit work flow. These displays can generate an increase in work efficiency and effectiveness. Visual displays also eliminate many work frustrations and help make the job more satisfying.

ERROR PROOFING

Error proofing is a Level 6 activity—the top of the visual factory pyramid. Error proofing is a process improvement that makes clear the occurrence of defects, equipment abnormalities, and safety hazards for fast analysis and solution. Error proofing prevents defects, abnormalities, and hazards from causing further defects, breakdowns, or injuries. In some situations, it permanently eliminates the possibility of defects, breakdowns, or injuries.

Error proofing uses physical hardware to eliminate or reduce defects. Even operational or procedural processes use some form of hardware to maximize performance. The parts most commonly used to error-proof manufacturing processes include electrical controls that sound alarms, trigger corrective measures, or temporarily stop the operation if an error is detected. Purely mechanical forms of error proofing include alignment pins, part-profile gates, and other part-interference devices. These provide a simple and cost-effective means of identifying, positioning, or blocking parts as they pass through the manufacturing process.

Simplicity is the key concept that makes error proofing effective. Most forms of error proofing are neither electrically nor mechanically complex. Some devices, however, take the form of elaborate electromechanical devices or PC-governed feedback/corrective action devices. See Figure 11-9 for some examples of er-

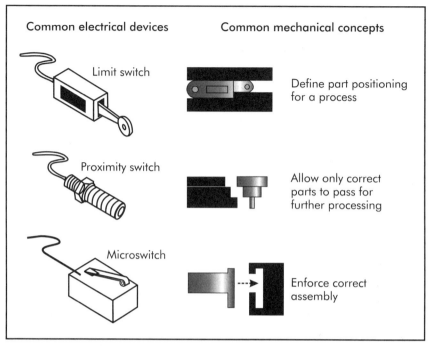

Figure 11-9. Examples of error-proofing devices.

ror-proofing devices. For a detailed discussion on error proofing, see Chapter 12.

IMPLEMENTING A VISUAL FACTORY SYSTEM

Create a visual factory implementation plan and assign responsibilities, then implement the visual factory techniques. To create a visual factory system, a straightforward implementation of the steps and levels described in the visual factory pyramid is needed. First, identify and prepare the area to be implemented. Then, choose a starting point within the production area. The starting point is not as important as participation from all parties involved in the process.

Once the area has been identified and a starting point has been selected, follow the 5-S strategy of workplace organization discussed earlier.

After organizing the workplace, implement visual controls and displays to support clear communication of the production standards and to signal problems and abnormalities for a quick response. Review and select the proper visual factory techniques to be used. Conduct periodic reviews of the visual factory's current condition and improve techniques where possible. Standardize any changes implemented in the visual factory system and clearly communicate the desired standards.

CONCLUSION

Implementing a visual factory system can help the lean implementer in many ways. It can:

- reduce the amount of time spent searching for supplies, tools, equipment, people, or information;
- eliminate many frustrations, such as those caused by not having the supplies, equipment, or information where and when it is needed;
- increase safety in the workplace by warning of safety hazards, communicating safety standards, and eliminating obstacles and unsafe conditions;
- improve communication between co-workers by standardizing certain types of communication mechanisms (such as check sheets, status boards, signs, and labels); create a common understanding about how operations should be performed; provide necessary information in a usable form at the point of use, and
- increase job satisfaction by producing all the results listed above and revitalizing the workplace through worker participation.

Error Proofing   12

By Richard Dixon

The primary reason to error-proof processes is to ensure safety in various applications. The secondary reason is to protect existing business by providing the quality levels that customers expect, and at a price they are willing to pay. The old business rule of "cost plus margin equals selling price" has changed to "selling price minus cost equals margin." The selling price is identified by the customer in today's globally competitive environment. To maintain or increase profits, everyone must focus on eliminating waste to reduce cost. Error proofing is an effective method of reducing costs through defect elimination. Achieving zero defects is the ultimate quality strategy of every organization.

Error proofing is one of the key ingredients to achieving stability—it is the first stop on the road to becoming a lean company. Lean manufacturing strives to continually eliminate waste; therefore, error proofing must reduce waste in the process. Waste is reduced by decreasing or eliminating defects, repairs, and scrap. All of the associated items, such as material handling, storage locations for the defective product, space to repair the product, and space to temporarily store the scrap, are reduced as well. Other items that directly benefit from error proofing include improved profitability and a reduction in:

- late deliveries,
- overtime,
- throughput time, and
- workers' frustration levels.

Error-proofing processes must be stable and predictable prior to moving to the second stage of lean implementation—continuous flow. (For a full discussion on the phases of lean implementation, see Chapter 7.) Stability includes making sure the environment has been cleaned; unnecessary items are disposed; everything that remains is located in its designated place; preventive maintenance is practiced religiously; and process variations have been minimized. Error proofing reduces variation, which in turn increases predictability.

The following comments may be heard in environments where error proofing is not effectively practiced:

- "We lost the contract because our quality was bad."
- "We don't have room to bring that work back in house."
- "My material handlers work overtime every week."
- "We have to start an emergency changeover—the parts in assembly are no good."
- "I need a Butler building to store rejected material until we can get time for repairs."
- "I need a Butler building to set up a repair area."
- "We missed a shipment—it was pulled from the assembly schedule because a component was rejected."
- "We have to work this weekend to remake the defective part."
- "We can't reduce our lead times—too many parts go through repair operations."
- "Our schedules would be more consistent if our product quality improved."
- "The increase in defects, repairs, and scrap caused us to miss our profit projection—no bonus, again."

Error proofing a process means that a method of intervention is placed in that process to either find defects or prevent them from being passed on. Prevention is preferred because it eliminates the cost of rework, carrying defective inventory, and moving the defective product in and out of repair operations.

Defective components can be segregated by shunting them to a rework slide. This is a detection error-proofing device; it detects the missing operation or component and shunts the part to a repair slide.

A prevention error-proofing device prohibits the production of defective parts. For example, by adding a locating pin to the fixture and a corresponding locating hole to the part, a symmetrical part can no longer be incorrectly located.

DEFECTS VERSUS ERRORS

When defining the concept of error proofing it is important to distinguish between defects and errors. A defect is a deviation from the specification and results in the product failing to meet the customer's expectations. An error is a deviation from the intended process. All defects are created by errors, but not all errors result in defects. The following are some common errors in a manufacturing operation:

- Processing errors are caused by an omission and improper sequencing. For example, a part may receive too many or not enough holes during a punching process, resulting in a defect.
- During the assembly process a component may be omitted, which results in a defective product that requires either rework or is scrapped.
- If a part is incorrectly cut, or punched, it becomes a wrong workpiece later in the production process and becomes scrap.
- Using the wrong press to form a part can result in an operations error. The part may not form correctly and most likely results in scrap.
- Improper machine adjustments can produce defective parts. Human measuring errors can lead to unnecessary adjustments to the operation, which result in defective products.
- Errors in equipment maintenance or repair occur when equipment is not properly maintained or repaired, resulting in defective parts.
- Errors in preparation of blades, jigs, or tools can occur during setup and lead to defective parts.

WHERE TO APPLY ERROR PROOFING

How does a lean implementer or project manager identify processes that need error proofing? Study the operational data. Review a Pareto analysis of defects for a given plant, department, or process, and the data will lead to the correct path. Effort and energy should be expended on the top three defects. It is important to realize that all processes can be error proofed. Focusing on the 10th priority item in a Pareto chart will reduce or eliminate that particular defect, but will not have the same impact on the operation as solving the most significant defect. If a situation presents limited resources, focus on the top items in the Pareto chart. At the same time, however, begin teaching the error-proofing process to everyone in the facility. Every operation can be improved, and each person can become an error-proofing engineer. Which makes more sense, using four engineers for error proofing or using 500 operators? When the entire operation is submitting ideas, maintenance resources will be swamped. It is best to roll out error proofing in stages (by group, area, department, etc.) to give maintenance personnel time to work on solutions. The best approach enables the operator to create or help create the error-proofing device.

Error proofing should not be expensive. Part of the training process is to teach employees how to come up with simple, inexpensive solutions to reduce and/or eliminate defects. They know the defects their operations are creating, and they probably have good ideas about how to correct the problems inexpensively.

Recognizing Error-proofing Opportunities

Red flag conditions identify areas where problems occur most frequently. As the project manager studies the following red flag conditions, he or she should visualize the processes in which they exist. Each of the following red flag conditions can be found in nearly every manufacturing plant. When the condition is recognized, error proofing it reduces defects.

Adjustments

Most equipment in a manufacturing facility requires periodic adjustment. Grinders, shears, presses, notchers, broaches, and other types of equipment require the operator to make adjustments to ensure that quality parts are produced. Operations that require frequent adjustments run the risk of an incorrect adjustment that produces defects. Error proofing these operations produces a fast payback.

Tooling/Tooling Change

Stamping press operators can make the mistake of selecting an incorrect die pair when the dies are similar, even with identification stencils. Color-coding is a good visual method of error proofing for dies.

Critical Dimensions or Conditions, Narrow Specifications

Characteristics such as critical dimensions or conditions and narrow specifications can be controlled with error proofing. For example, when finishing a paint process, the conveyor speed and oven temperature combine to cure the paint finish on the vehicle. If either of these critical components are not controlled, the body will not cure properly and result in rework or scrap.

Many/Mixed Parts

Operations that assemble kits are always at risk of shipping a kit minus one or more components. Error proofing the kitting operations results in happy customers and reduced cost. In assembly operations, managers ensure that all components are assembled as specified by using effective error-proofing techniques such as color-coding screw heads.

Multiple Steps

Many parts go through multiple operations during the manufacturing process. The fact that multiple operations are required increases the probability that a particular operation can be omitted. Cellular operations and transfer equipment also make it possible that an operation could be missed. Error proofing can prevent omissions caused by multiple-step processes.

Ineffective or Nonexistent Standards

Ineffective or nonexistent standards cause operators to make unguided decisions during the production process, which can introduce variations into the process. Variation results in errors, which produce defects. All operations should have standardized work instruction (see Chapter 13) visible at the process. Containers should always have a standardized method of place-packing and a standardized quantity so the space can be used most efficiently and avoid operator errors. An orderly packing process may free up previously occupied space for another use.

Infrequent Production

Parts or operations that run infrequently often can produce poor quality results. The tooling is usually difficult to find, the setup process is unfamiliar, and operators have relatively little experience running the parts. Tooling should always be stored in its designated location. Error proofing the tooling, using locating pins or similar devices, helps ensure that the setup occurs correctly. Following standardized work instruction helps produce a quality part.

Symmetry

It is possible that a part with symmetrical edges or protrusions can be incorrectly installed. Figure 12-1 shows the difference between symmetric and asymmetric tabs during the assembly process. Increasing the width of one of the tabs removes the possibility of incorrect assembly.

Asymmetry

In some product structures, a part is identical except for one dimension. For example, the steering rod shown in Figure 12-2 has a ground finish on one end and a broached finish on the other. The end of the rod with the ground finish can be inserted into the next process incorrectly, while it is impossible for the broached end to be used improperly.

Rapid Repetition

Defects or errors in high-speed equipment can quickly make an enormous pile of rework or scrap. A prevention sensor in the au-

Error Proofing

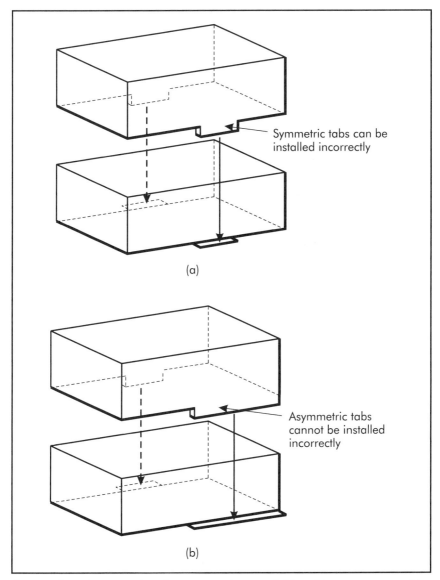

Figure 12-1. Asymmetric tabs remove the possibility of incorrect assembly.

tomatic welding process would prevent defects from moving to the next operation. If installing a prevention device is not a viable alternative, a detection device should be used. Effective error-proofing detection for this condition may be a simple bar device

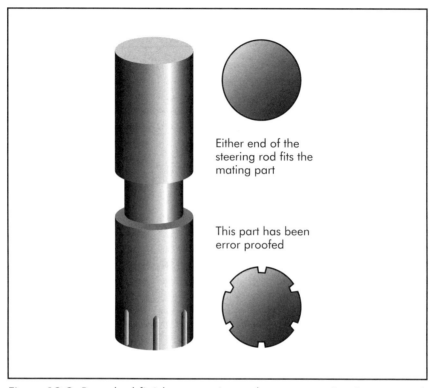

Figure 12-2. Broached finish on steering rod ensures against incorrect orientation in the next process.

that separates good parts and moves defective parts into a rework bin. In some cases, technology does not exist to prevent defects. In many cases, cost cannot be justified for prevention devices, so inexpensive detection devices are used.

High-volume Production

High-volume operations are similar to rapid repetition operations. Rapid repetition operations run quickly in low- to medium-run quantities. High-volume operations run for long periods making the same part. The quality risk in high-volume operations is tool wear. Sensor monitoring of tool wear is more effective than operator inspection on a periodic basis because the variety of materials processed results in nonstandard tool wear.

Cycle Time Exceeds Takt Time

When customers require products faster than they can be made, cycle time exceeds Takt time. This condition results in a greater number of defects because of the extreme pressure to deliver products on time. Error-proofing processes in this situation result in less scrap and rework, which reduces the chance of shipping the customer defective products. It also results in less overtime costs. Efforts should be made to reduce the cycle time to below the Takt time through workplace organization, preventive maintenance, and quick changeover. If these lean methods fail to reduce the cycle time below the Takt time, then additional capacity should be purchased.

Environmental Conditions

Environmental conditions can result in a red flag, and overlooking the cleaning operation is a common error. Stamping operations often have problems with die lubricant overspray. Parts leave with excessive oil and often have to be cleaned before shipment, resulting in extra operations, extra transportation, and additional cost.

PHASES OF ERROR PROOFING

Figure 12-3 illustrates the progression of error-proofing techniques. The left side represents a lack of error proofing. As the company moves upward toward prevention, it passes through the human judgment and detection phases of error proofing. The human judgment phase relies heavily on the operator responding to a potential error. The detection phase focuses on the installation of detection devices. During the prevention phase, the focus is on product and process design, which eliminates the possibility of errors.

Many companies conduct intensive lean design training in the early stages of lean implementation to ensure that design issues do not lead to quality problems during future production processes. The results of a company's success in error proofing can be found in warranty costs, number of defects supplied to the customer, and quality Pareto charts used in the production process.

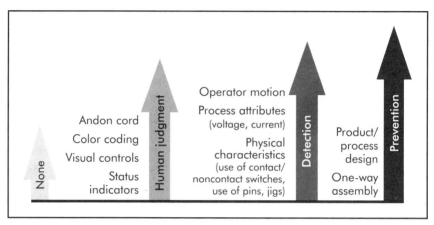

Figure 12-3. Hierarchy of error-proofing techniques.

Equipment and process designers have the primary responsibility for implementing error proofing into every new process. Thorough reviews of the product and processes with process operators must be conducted prior to machine/process design to ensure that error-proofing opportunities are not overlooked. Design engineers must grasp the situation on the floor. Conference room meetings may resolve some of the issues, but engineers and operators must review the equipment and processes at their physical location. Limiting error-proofing meetings to the conference room results in missed opportunities.

For existing equipment/processes, operators have the primary responsibility of identifying error-proofing opportunities that are justified by quality and downtime data tracked with Pareto charts. Once opportunities are identified, assistance from engineering or maintenance may be required. The focus should be on no-cost or low-cost forms of error proofing.

Checking Methods

Nearly all production processes can be error proofed based upon one of the following three checking methods:

1. In the feature method, a physical characteristic of a part is sensed to differentiate it from the standard for that charac-

teristic. In the design stage, standards should be established for physical characteristics of the product and process operation. Sensors, jigs, and locating pins are effective error-proofing devices for physical characteristics. The most effective sensors are limit, proximity, microswitch, fiber, and photoelectric.
2. In the constant value method, either a predetermined number of motions are performed by the operator or the machine maintains control within specified ranges. Light curtains and switches are effective error-proofing methods for operator motions. For machine control, error-proofing devices that monitor torque, angle, air pressure, weld current or voltage, flow, level, temperature, time, or dimensional characteristics are most effective. When a characteristic falls out of the minimum/maximum range, the machine should shut down.
3. The motion step method consists of a distinct sequence of operator motions or machine processes. Standardized work instruction offers the best method of controlling operator motion by making sure the motion sequence is formed consistently and without variation. Cell design ensures process sequence flow.

The sensor chart shown in Table 12-1 identifies a number of sensors and their effectiveness in error proofing the characteristics listed.

STEPS TO EFFECTIVE ERROR PROOFING

There are nine steps to effective error proofing:

1. Identify and describe the defect or red flag condition in detail with supporting data. The red flag conditions identified earlier in this chapter are a good place to begin. Use data to identify which part of the process should be error proofed first.
2. Gather historical trend data over a significant period of time. Current data may not reflect the true problem. Review historical data to make sure there is a correlation.

Table 12-1. Error-proofing sensors

Sensor Type	Lack of Parts	Number of Revolutions	Color	Dimensions	Shape	Thickness	Safe-guards	Presence/Absence	Material
Inductive proximity		XX			XX			XXX	
Through beam								XXX	
Retro-reflective beam								XXX	
Diffuse reflective beam			XX					XXX	
Fiber-optics through beam		XX						XXX	
Fiber-optics reflective beam		XX	XX					XXX	
Fiber-optics heat-resistant		XX	XX					XXX	
Micro-photo switch		XX						XXX	
Ultrasonic wave								X	
Analog proximity		XX		XX		XXX		X	XXX
Area							X	XXX	
Displacement				XXX		XXX		XXX	
Size recognition				XX		XXX		X	
Analog beam			XXX					XXX	
Press safeguard							XXX	XXX	
Image	XXX			XXX				XXX	
Color visual	XXX		XXX					XXX	

XXX = Most effective   XX = Somewhat effective   X = Least effective   Empty cell = sensor cannot be used

3. Brainstorm potential reasons for the defects. The brainstorming session should include operators, maintenance people, supervisors, and engineers. Remember, there are no dumb ideas. Use data to support each potential reason.
4. Select the most likely cause and use data to support consensus decision-making.
5. Develop a problem statement. Make sure the statement has a subject, verb, and object. The object is the problem's impact on the operation. The problem statement must not suggest a solution, or the "5-Why" analysis will not get to the root cause.
6. Conduct a "5-Why" analysis. Use the five "whys" (maybe even seven or eight) to find the root cause of the problem. When that is completed, begin with the last "why" statement and use the word "therefore" to answer it. This process determines whether the analysis makes sense. Continue this process through the preceding "why" steps. The process should yield logical statements—if not, it needs to be repeated.
7. Identify a design solution. This step identifies the who, what, when, where, why, and how. Use people's names instead of departments, and get their agreement before listing them as the responsible party for an action item.
8. Determine the cost justification. Remember that most error-proofing ideas can be implemented with little or no expense. If that is not the case with the design solution, return to step 7 and develop a low-cost solution if possible. This step is critical for obtaining upper management's approval. Identify the hard savings that the error-proofing project will generate. Soft savings (such as square footage or worker's compensation projections) generally do not carry much weight in the approval process.
9. Implement the solution. If the process has been followed, the solution will eliminate the defect. If so, congratulate the team members and move on to the next opportunity. If the process was not followed, return to step three and start again.

CONCLUSION

Error proofing works best when it is designed into a process for delivery to the workplace. If that is not possible, the concepts and practices addressed in this chapter lead to effective error proofing of existing processes. Error-proofing design is based on firsthand experience, observing processes, and knowing the errors that can occur. Observations and data drive the error-proofing process, so process designers must grasp the current situation on the plant floor. If error proofing is introduced into the current process, it is critical to involve production workers.

Figure 12-4 lays out the hierarchy of quality inspection effectiveness for zero-defect strategies. The first four levels in the chart result in defects—only level five results in defect-free production. To ultimately survive in today's environment, customers must be supplied with defect-free products. Error proofing will make that a reality.

Error Proofing

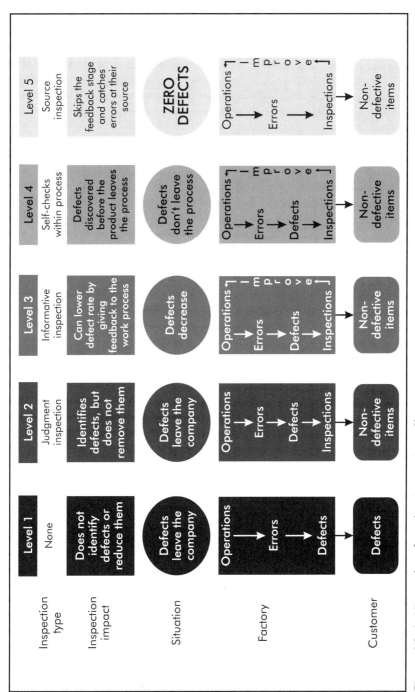

Figure 12-4. Hierarchy of quality inspection effectiveness.

Standardized Work 13

By Gregory Thomerson

Standardized work is a method used by the operator to organize his or her tasks in a safe and efficient manner. At the heart of standardized work is the organization and specification of uniform work steps to be performed in a manufacturing process. These steps are documented in a standardized work sheet that is displayed in a prominent place at the workstation. Standardized work documentation should be used anywhere there is a definable and repeatable process.

Standardized work is a process owned by the operators. Engineers do not develop the work documents; rather, they are created, owned, and maintained by the operators doing the work, and are displayed near the process or the machine on the line. This chapter describes the purpose and advantages of the standardized work method in the lean manufacturing environment.

BENEFITS OF STANDARDIZED WORK

Standardized work reduces waste by identifying and eliminating unnecessary motion and effort. It outlines safe and efficient work methods. When standardized work processes are followed, they maintain quality and prevent equipment damage. Standardized work serves as a foundation, or baseline, for improvement and provides operators a means to define their jobs. In addition, it creates consistency between operators and shifts.

Other benefits of standardized work are as follows:

- There is a safe method for work. Standardized work sheets spotlight potential hazards and the required safety steps.

- Safety procedures are defined and displayed at the workstation to ensure that safety is maximized for the operators.
- It provides a predictable, stable work pace. Work processes are defined and documented, providing a predictable work sequence. This gives operators security in knowing how to complete their operations. When the steps are defined and repeated, there is less chance for accidents to occur.
- It prevents overproduction, an important source of waste. Customer demand determines production quantities, and the workload is determined from the production rate needed to satisfy customer demand. This also reduces excess inventory.
- It stabilizes, maintains, and controls quality. Operators are taught using standardized work instructions to ensure consistent training and eliminate errors. Since operators' movements are clearly defined in advance, they no longer have to constantly determine the next process step.
- There is a stable platform for continuous improvement. Criteria are set for normal and abnormal production during operation. A controlled environment is established, allowing the effects of change to be visible.
- It leads to lower costs because consumption of in-process stock is more predictable. Actual process improvements can be captured and documented. Stable and accurate work reduces the possibility of machine breakdown. Uniform distribution of work results in a stable workload.
- It helps adjust Takt time and line speed. Adjustments are easy to make when standardized work is implemented. Takt time can be adjusted to the present order. When the production schedule changes, operators can look at the posted standardized work sheets to determine how people should be redistributed within the plant. This determines the best arrangement of operators and resources.

PURPOSE OF STANDARDIZED WORK

Operating abnormalities and waste are easily identified because standardized work clearly shows the current method of operation. In production operations, standardized work helps maintain consistent quality, provides safe, efficient operations, and ensures the proper use of equipment.

Standardized work serves as a visual management aid for the supervisor. As an auditing instrument, the standardized work document should be monitored and compared with the operator's movements and work elements. Variations or deviations must not continue without being addressed. The difference between work being completed and work elements on a standardized document should be pointed out. The difference should be noted and the need for continuous improvement reinforced.

Continuous improvement activities rely on the foundation of standardized work. It organizes and defines the operator's movements. Without standardized work instruction, worker movements are disorganized, and establishing a baseline to document the current condition is nearly impossible. Also, it is paramount that movements and work elements are organized when training new operators. Organizing work elements eliminates confusion and frustration for the operator.

Standardized work does not remain fixed. As new requirements or standards are implemented, the operator must be aware of the changes and assume a role in integrating the new information into standardized work instruction. Operators must put changes into practice and revise the standardized-work document. Standardized-work documents are frequently reviewed and revised as long as there are changes and improvements being made. The document should remain posted in the workplace at the point of use.

ELEMENTS OF STANDARDIZED WORK

Standardized work sheets provide a visual control that ensures consistency. Work sheets are a set of papers that document work processes, provide a visual control, and can be used as a training tool. Work tasks are defined on standardized work forms and are separated into basic elements that can be structured and defined. The basic elements of the work sequence are:

- Work steps—elemental descriptions of the work needed to complete the tasks in each process.

- Safety steps—all safety checks, precautions, and equipment are defined on the standardized work forms.
- Visual representation—layout of work and processes that take place at the workstation are defined on the standardized work forms. These sequences are visually defined to provide a visual control.
- Timing—time spent at the workstation is separated into walking time, manual work time, and automatic work time.
- Takt time—the total available operating time divided by the number of units required by the customer.

Work Step Elements

Identifying the work elements of a repeatable process is a critical part of developing the standardized work sheet. When preparing for documenting, developing, and implementing standardized work, managers must capture the current conditions of the process operation. Begin by identifying movements and steps currently performed in the operation, especially the major steps.

A *work element* is a small, natural grouping of actions performed by an operator, such as joining a bracket to a plate. Elements are the basic building blocks of standardized work. There are three basic types of work elements:

- Work—actions relating to assembling, machining, processing, etc., which are generally value-added activities.
- Walk—movement from one place to another.
- Pick—motions necessary to reach for an object, gain control of it (for example, close fingers around an object), and withdraw or move the object from its surroundings.

The following are some examples of work elements:

- A bracket is assembled to a plate using three bolts and tightened with a torque gun. The bracket does not stay in place when only one bolt is installed. Therefore, it makes sense to hand-start the first bolt, install the second bolt with an air gun, install the third bolt, then tighten the first bolt. It is not appropriate to treat each bolt as a separate element. Separating them would require repeated handling and inefficiency.

- An assembler is responsible for putting in the roof liner of a vehicle. As part of the process, he or she installs a piece of molding so the roof liner stays in place without falling. If this piece of molding were not put in, the roof liner would fall. Both activities should be listed as one element.
- An operator picks up a bracket, places it on a fixture in a nut welder, and hits the cycle button. It would not be efficient to break these actions up among several operators. It is better to group these actions together as one element.

As elements are identified, elements can, in turn, be grouped together to develop an effective work sequence with minimum waste. However, when developing a work sequence, it is important not to put elements together that create ergonomic problems for the operator (for example, extended periods of bending, overhead work, etc.).

When writing elements on a standardized work sheet, they should be phrased using action words, such as install, press, place, pick up, connect, or load.

Timing Work Elements

Measuring the time of work elements has two basic benefits. It recognizes each element and acknowledges that each step has a target or standard time allocated to it. This makes it easy for the operator to understand each element and become familiar with the required action. The time measurement of each element does not necessarily have to be displayed on the standardized work sheet.

A key part of standardized work is capturing the time required for each element. To take this measurement, first determine its starting point. For example, the beginning of a work element is typically distinguished by one of the following actions:

- Hearing (or listening for) when the work element begins. Sometimes the starting point may be recognized when hearing the sound of a piece of equipment or hand tool (such as an air gun on a bolt or nut).
- Observing when the operator picks up a tool or touches a part. This is the most common method used to determine the starting point of a work element.

- Observing footsteps made by the operator to get properly positioned in the process layout.

STANDARDIZED WORK DOCUMENTS

Three forms make up the standardized work documentation—the standardized work sheet, the standardized work combination sheet, and the production capacity sheet. Together, they represent the standardized work process. The forms are interrelated and their use depends upon the situation.

Standardized Work Sheet

The *standardized work sheet* provides a broad overview of workstation layout and work sequence at a particular operation. It documents items such as Takt time and the sequence of work elements, and identifies elements designed for self-monitoring and operator ownership of the process (quality, safety, and critical processes). It also includes a diagram that visually displays the work area, safety procedures, and quality checks. An example of a standardized work sheet is shown in Figure 13-1.

Standardized Work Combination Sheet

The *standardized work combination sheet* graphically represents operation time, which enables the work sequence to be analyzed and identifies wasted time. It combines manual work with machine processing time to ensure that both can be performed within Takt time. The sheet is a tool to determine the range of work and sequence for which an employee is responsible. It also can be used as a tool to analyze the relationship of the operator and equipment. An example of a standardized work combination sheet is shown in Figure 13-2.

Production Capacity Sheet

A *production capacity sheet* shows the production capacity of the process and documents the time required for manual and

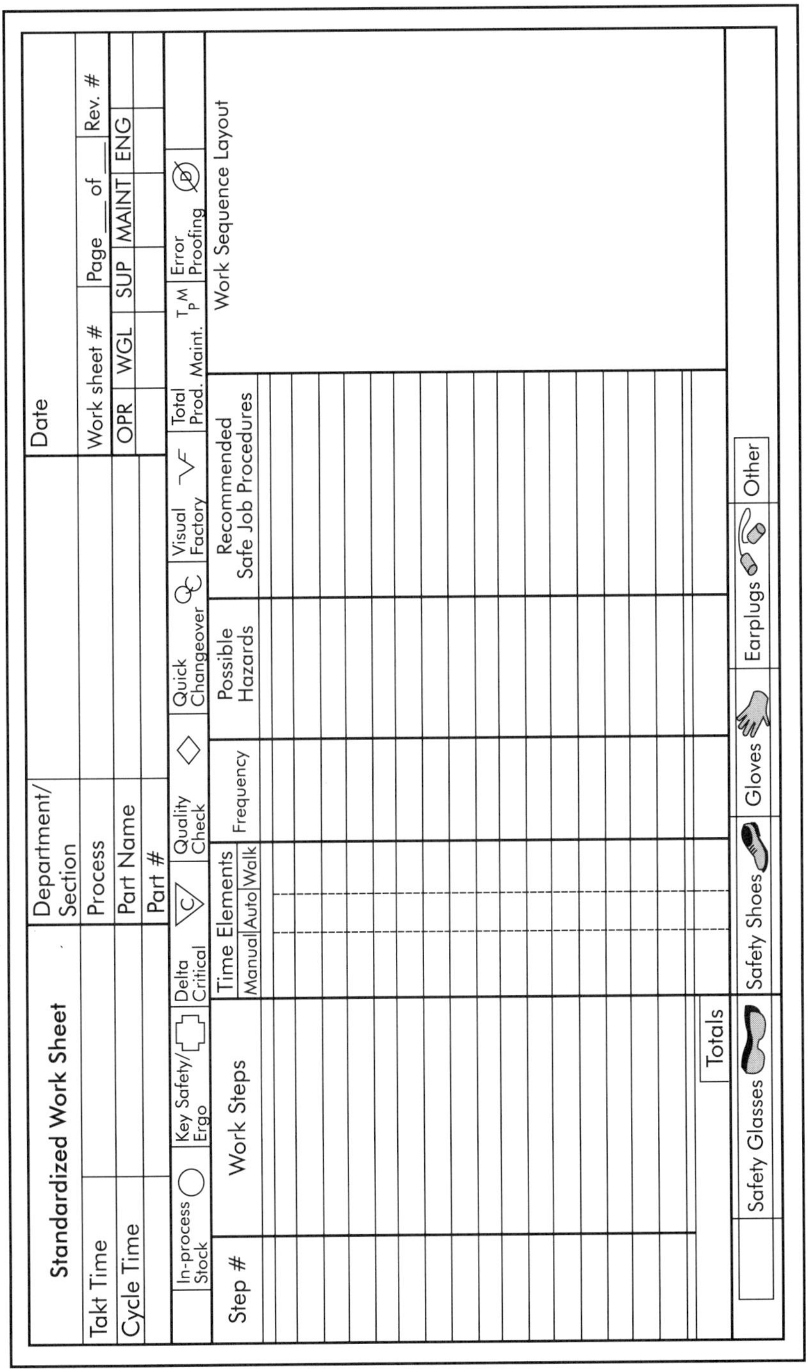
Standardized Work Sheet
Department/ Section
Date
Takt Time
Process
Work sheet #
Page ___ of ___
Rev. #
Cycle Time
Part Name
OPR | WGL | SUP | MAINT | ENG
Part #
In-process Stock | Key Safety/ Ergo | Delta Critical | Quality Check | Quick Changeover | Visual Factory | Total Prod. Maint. | Error Proofing

Step #	Work Steps	Time Elements: Manual	Auto	Walk	Frequency	Possible Hazards	Recommended Safe Job Procedures
	Totals						

Work Sequence Layout

Safety Glasses | Safety Shoes | Gloves | Earplugs | Other

Figure 13-1. Standardized work sheet.

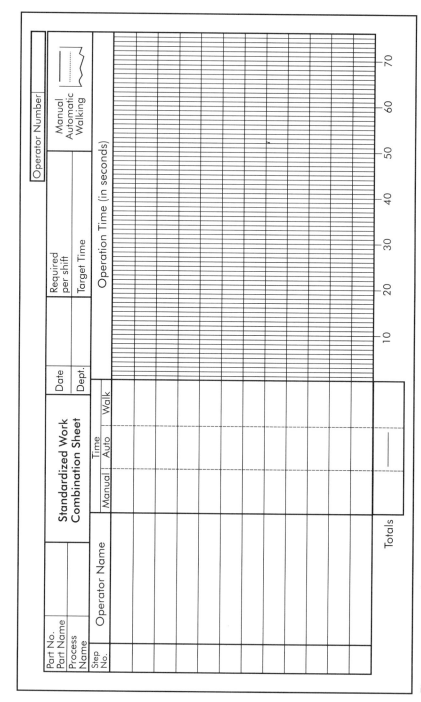

Figure 13-2. Standardized work combination sheet.

non-routine work. It can be used to locate problems or bottleneck processes along the line, and, over time, can be used to evaluate machine performance. An example of a production capacity sheet is shown in Figure 13-3.

Maintaining Standardized Work Documentation

Developing and updating documents should be the responsibility of those with the most knowledge of the jobs in question. The team leader often drives this process. Supervisors and managers are responsible for ensuring that documentation is posted and updated, and that operators are performing the work according to the standards. The standardized work system includes a regular monitoring schedule and audit practice. This is designed not to threaten the operator, but to reinforce the principle that standardized work allows and encourages change and continuous improvement.

PRECONDITIONS FOR STANDARDIZED WORK

As attempts are made to implement standardized work, a number of conditions can impede standardization. A few basic requirements must be met when assessing the readiness for standardized work procedures.

Operators must have a safe, clean, neat, and orderly workplace with a specific location for everything. Anything that is not required for the job must be eliminated. Good workplace organization identifies waste such as overprocessing, conveyance, excessive work-in-process inventory, motion, and waiting. Operators must stabilize their workplace by eliminating unnecessary items, having fixed locations for tools, equipment, and parts, and maintaining cleanliness. Stability enables repeatability and predictability, and is vital for standardized work elements to begin. See Chapter 11 for more on workplace organization.

Observe the operators on each shift and determine which movements provide the least amount of waste, overburden, or unevenness. These movements become the baseline for the standardized work process. Managers then can require that each operator use the best practice.

Group: Name:

Date (new/rev.):___/___/___ Revision #: Page ___ of ___

Manager	Supervisor		Productivity Capacity Sheet	Part No.	Line Name		Current Output Person-Day
				Part Name	Max. Output Daily	Workers Needed	PCS

Step No.	Process Name	Machine No.	Walk Time	Base Time			Tools			Processing Capacity	Remarks	Time Indicated (Manual ——— / Automatic - - - - / Walking ~~~)
				Manual Time	Machine Time	Time to Complete	# of Pieces to Change	Time to Change	Time/ Piece			
		Totals										

Figure 13-3. Production capacity sheet.

Establish a consistent work sequence for all operators based on the best sequence used by those familiar with the operation. If movement or work sequences vary each cycle (between operators or shifts), it is difficult to even attempt to standardize it. Humans are creatures of habit, and most operators perform the production process sequence the same way each time. Standardized work captures that current best method and documents the sequence.

Equipment uptime must be optimized and breakdowns must be minimized. Equipment and machinery breakdowns interrupt the flow of any operation. Interruptions cause operators to become frustrated and overburdened when performing difficult tasks, out-of-standard methods, or waiting for repairs. The rhythms of the operator's movements become ragged and jerky instead of smooth and flowing.

Another consequence of poorly maintained equipment is questionable quality. Stable and consistent quality is vital for standardized work. If the quality of the incoming part is unstable, operating conditions are constantly changing. Every time a quality problem occurs (either at the current or prior operation), or there is a variation in precision, an investigation should be conducted and countermeasures or changes put in place. As a result, operators may have to adjust their work method or use more time to complete their work. Wherever this happens, repetitive work becomes unstable and standardized work will not function as intended.

ROLES AND RESPONSIBILITIES

Establishing standardized work procedures during the lean implementation process is important for many reasons. Foremost is the way ownership shifts work from outside the process to the operators. The standardized work process provides operators with a way to control their jobs, which allows them to contribute to productivity improvements and see the results of their efforts. The standardized work process is founded on the belief that the operators doing the job have the best knowledge of how the work should be performed.

The philosophy of standardized work is aligned with the principles of lean manufacturing. Both entities are work-group centered, which means that operators are empowered to make decisions that impact the work group and overall production. Decisions and evaluations are no longer coming from an outside source—they are internal to the work group. Given the opportunity to make decisions that lead to improvements in product quality, operators can think creatively and find innovative solutions to production problems.

For standardized work to happen as intended, members of the production team must fulfill their roles. For the operator, this involves defining work steps and completing the standardized work forms. The operator also must update forms after any continuous improvement activities and maintain the forms regularly.

There are a number of actions managers can take to ensure success with standardized work:

- Recognize when deficiencies or abnormalities occur. Standardized work becomes the main visual control element used by everyone to monitor and identify variances from the desired or standard expectation.
- Make a commitment to lead. Take the initiative to gain a personal understanding of the process, and to maintain and improve the current situation. Commitment means going to the production floor and observing standardized work with the operators and the production supervisor.
- Managers are responsible for allocating the time needed for training the entire organization about standardized work. They must provide support within the operation and develop a procedure to ensure that everyone enhances their skills to perform standardized work.
- Managers must demonstrate their own knowledge of the standardized work system, act as a coach, and provide operators with time to practice their newfound knowledge. In addition, it is important to continually support standardized work by exercising its use of visual control. It is beneficial to encourage and lead discussions and interaction with production process operators. Visiting the workplace frequently and personally confirming the use of and adherence to standardized work is encouraging and reinforcing.
- Provide coaching and encouragement to production personnel and assure operators that they have ownership of the

process. Demonstrate that it is okay to have problems as long as they are quickly identified and steps are taken to get back to the standard. Deviations from the known expectation must be followed up with process personnel.

CONCLUSION

The goal for implementing standardized work should be that it is understood and followed by all employees. Once standardized work is established and understood, it is important to visibly place standardized work sheets at every workstation. Operators then can easily consult the forms for instructions and assistance. By using standardized work, everyone's work becomes more focused, more productive, and safer.

Quick Changeover 14

By Rod Centers and Steve West

Some people define *quick changeover* as a process used to continually reduce setup and changeover times within a plant. Others describe it as a system of methods and procedures to shorten the time it takes to change parts, or as a tool to eliminate waste within changeover processes. All three definitions have merit. Quick changeover's primary benefit is that it shortens changeover and setup times by using continuous improvement to attack and eliminate waste associated with current methods. Basically, it reduces the time the machine is not running. It can have a great impact on manufacturing companies if used properly. This chapter emphasizes how and why quick changeover techniques are accomplished.

HOW MUCH TIME REDUCTION IS POSSIBLE?

Quick changeover is by no means rocket science. It is a simple, hands-on tool that is easy to understand and implement. Some companies have changeovers that take hours, days, and, in some cases, weeks. Many changeovers do require a lot of time due to cooling procedures and the number of machines involved. However, any company that performs changeovers can reduce waste.

Many companies today make improvements, then become complacent. The question that begs to be asked is, "how good is good enough?" The answer is that it is never good enough. Is the competition making improvements? Is the competition satisfied with the status quo? In today's world, there is someone out there working hard, figuring out how to satisfy the customer a little more than your company does. They are working on reducing cost to offer the same product a little cheaper.

Several years ago, companies used a simple formula to determine the selling price of their products. They simply took their costs, added the profit they wanted to make, and that was their selling price. In today's competitive market, that formula no longer works. The customer is now setting the selling price. With competition as tough as it is, and with more companies competing for the same market share, the company offering the lowest cost, highest quality, and best delivery gets the business. The only opportunity to increase profitability is to reduce production costs. Managers must reduce costs if the company is to remain competitive. Quick changeover is one tool that will greatly assist their efforts.

HOW QUICK CHANGEOVER HELPS

By reducing the time it takes to change parts or tooling sets, any organization will quickly realize impressive benefits. In today's extremely competitive manufacturing environment, companies must strive to reduce waste and improve quality and profitability if they expect to survive in the future. The following sections give examples of how quick changeover can help an organization.

Increased Availability of Equipment

Increased equipment availability translates into more products being produced. Companies will see measurable improvements due to:

- overtime reduction (by becoming more efficient and responsive to customer demands, overtime hours can be reduced or eliminated);
- possible elimination of afternoon or midnight shifts (as processes become increasingly efficient, it takes fewer man-hours to produce the required amount of product);
- better operator efficiency (as changeover times decrease, equipment operators gain efficiency through machine availability and batch-size reduction), and

- more parts per person (with fewer man-hours needed to produce the required amount of product, total costs decrease and revenues increase).

Better Response to Customer Demand or Schedule Changes

In the manufacturing world, scheduling changes are a constant struggle. Often, schedule changes result in excessive overtime, higher freight costs, and customer dissatisfaction because of missed shipments. By reducing changeover times, managers can better meet changing schedules within established operating patterns. Increasing flexibility generally leads to a few different results:

- There is better shipping schedule attainment. Managers can more effectively meet shipping requirements when time wasted by changing parts is reduced. Since the item is produced more efficiently, it can be shipped to the customer on a more predictable, stable basis.
- There are lower premium freight costs. Since the product is being created more efficiently and is shipped on a more stable basis, premium freight costs are drastically reduced.
- The company enjoys higher customer satisfaction. The customer cares only that the company delivers quality products, within cost, and when desired. As changeover times are reduced, managers will quickly realize the benefit of meeting and exceeding customer expectations.

Ability to Reduce Lots or Batch Sizes

The ability to reduce lot sizes is one of the greatest improvements of quick changeover. Most companies produce in large batches for a variety of reasons, including the fear of equipment breaking down, perceived convenience, and excessive changeover loss. Lengthy changeover times force companies to make product in large quantities. If the management team knows it is going to lose seven or eight hours on a piece of equipment due to the changeover, it will try to reduce changeover times to as few as necessary. Following are some of the benefits of producing in smaller batch sizes:

- There is a reduction in inventory. Inventory is waste. Smaller batch sizes allow companies to produce to customer requirements and ship, rather than store, product.
- Less time is spent maintaining inventory. Many companies spend a great deal of time counting parts to ensure they have accurate inventory levels. Each minute spent maintaining inventory is absolute waste.
- Quality problems are quickly noticed. When goods are produced in large lots, they are stored until needed. Therefore, quality problems often are not noticed until a good deal of time has passed.
- There is a quicker response to customer demand. When customers demand changes, it is easier to shift to the new requirements when using smaller batch sizes.
- Material is handled more efficiently. With less inventory to move, the need for material handling decreases dramatically.
- Floor space is used more efficiently. Parts storage is not the best use for valuable floor space. How much money can be made from floor space used as storage compared with floor space used for production?

IMPLEMENTING QUICK CHANGEOVERS

Implementing quick changeover techniques requires managers to answer a number of questions: How in the world do I do it? Where do I start? How do I begin to make the needed improvements? How do I get the right people involved? How soon until I see results? The following are basic concepts and guidelines that implementers can use.

The Current State

The first step when implementing any kind of improvement is to understand where the company is (its current state). Managers must understand current methods and procedures, the current measurables associated with the task, and the amount of manpower spent completing the task. Without this level of understanding, improvement might take place, but how would one know?

Implementing improvements without knowing the current situation is like taking a trip without a destination.

Selecting an Area to Start

Choose an area within the plant that, if improved, enables the company to produce more products or drastically reduce costs. This area/improvement should have a high probability of success. When looking at the probability of success in a specific area, think about current methods, the willingness of the workforce in that area to make changes, and the leadership in the area and its relationship with the workforce. These factors can greatly influence the level of success.

Once an area is selected, document current practices and changeover techniques. This is a fairly simple step often forgotten or done haphazardly because many feel it is not a function for improvement. Documenting changeover provides an initial understanding of each operation so that improvement opportunities can be easily recognized. It is true that documenting current methods and procedures does not improve results. However, without knowing exactly where the company or process is and how it got there, improvements cannot be measured.

Documentation Tools

One method of aiding the documentation effort is to videotape the current changeover process. Videotape the entire process, including tooling/die removal, locating and retrieving new tooling/dies, locating new tooling/dies in machines, any set-up adjustments, quality control functions, etc. The video enables managers to view the entire process repeatedly and document the accurate timing of each step. Do not bother with trying to get everyone involved together when the changeover is taking place. However, it is strongly recommended that everyone on the plant floor participate in the quick changeover effort at some point. Since that is where the improvement is implemented, what better place to generate ideas and opportunities than where the physical work is taking place?

Another form of documentation is to simply record each changeover on the floor. Operators or skilled tradespeople can complete this item. Use records as tracking devices to monitor stability as well as long-term results within a process. Most companies develop some sort of documentation sheet for this purpose. See Table 14-1 for an example of a quick changeover documentation sheet.

Separating Internal and External Elements

With in-depth documentation of the changeover process in hand, managers can separate internal and external elements of the changeover process, then shift the internal elements to external wherever possible. *Internal elements* are performed while the machine is not running (for example, changing a die or tooling). *External elements* are performed while the machine is still running (for example, locating the tooling or die needed for the next production run). Much improvement can be obtained simply by transferring internal elements to external functions. Use changeover documentation to determine which elements are internal and which are external (see Figure 14-1).

Table 14-1. Quick changeover documentation sheet

Step Number	Description	Internal Time	External Time	Comments

	Changeover Activities	Internal	External
1	Scheduling the changeover		
2	Locating die and steel coil		
3	Removing old die		
4	Removing scrap from old run		
5	Aligning die placement in press bed		
6	Adjusting feed settings		
7	Performing quality control checks		
8	Informing material handlers of changeover and needed materials		
9	Setting height adjustments on press ram		
10	Acquiring needed dungaree for production run		

Figure 14-1. Determining the source of changeover activities.

Shifting Internal Elements to External

Once internal and external elements have been identified, the next step is to shift the internal activities externally. Remember that the primary goal of quick changeover is to reduce the time the machine is not running. Reducing internal elements helps shorten the amount of time the machine is idle. During this part of the process, managers must begin to think without regard for the past. Creativity and imagination must be injected here to ensure every opportunity is employed to reduce changeover times. Many improvement opportunities are not seen because current practices are accepted. Countless companies allow opportunities for improvement to pass by each day because they fail to understand that there may be better, more efficient ways of achieving the desired outcome.

By shifting the eligible internal elements to external, a typical improvement of 20–40% can be expected. The goal in this process is to analyze every internal element and determine if it is feasible to shift externally. Shifting a number of internal elements to external reduces the amount of downtime for the machine. Do not take anything for granted during this step. All elements should be considered. Simple, cost-effective ideas commonly create the greatest savings. Get input from all persons involved, from operator to operations support personnel to engineers.

Improving the performance of external elements while the machine is running can be one of the most effective and efficient ways to improve changeover. For example, locating tools, dies, and materials before shutting down the machine can pay big dividends in the changeover process. Make sure operators and skilled tradespeople are notified prior to the changeover so they can prepare what is needed. Once this has been done, begin to look for ways to shift internal elements to external.

One thing to consider when implementing quick changeover is presetting the gaging or tooling. Can settings be standardized to minimize set-up time? What about heights, locations, positions, and locking mechanisms? Can molds be preheated to the proper temperature before changeover? What is the best way to organize tools? These are merely a few of the hundreds of questions to be asked when trying to reduce changeover time. Once a machine is taken down, nothing should interrupt the flow of the tool change.

ADVANCE PREPARATION ILLUSTRATION

The following illustration of a lapping machine that uses multiple rolls of lapping paper is designed to help implementers think in terms of simple yet effective methods that are "out of the box." Let's use an example of advance preparations. What this means is simply getting the necessary parts (tools, etc.) ready before any internal activity begins. Conditions such as temperatures, pressures, height settings, and adjustments can sometimes be set while the machine is running.

The lapping machine may not stop or warn the operator until it is completely out of paper, which can be costly on a machine that runs a 30-second cycle time. For example, a machine goes down due to a lack of lapping material. The operator is busy on another operation and it takes him or her five minutes to respond to the downed machine. Meanwhile, the lapping paper required for setup is not beside the machine; it is in a tool crib three minutes away from the operation. What if there is a line of six people at the crib waiting for the single person working the crib since the other person who should be helping out is on vacation? As a result, the operator waits in line for seven minutes for the paper. The operator gets back to the machine (another three minutes) to do the change. (The assumption is the operator doesn't stop to talk, get a drink of water, or use the restroom.)

Now that the operator is ready to do the change, he or she finds out that the allen wrench is not where it was yesterday or the socket is missing because it was used on another operation this morning. The operator spends five more minutes locating necessary tools. When the operator eventually begins the changeover, it takes only 15 minutes instead of the usual 20. The operator is feeling good about the changeover and begins to run parts to perform a first-piece quality check. Oh no! The gage is missing. The operator spends four minutes looking for it. After realizing the gage cannot be found, the operator calls the gage room to find a replacement. After 15 more minutes (which would be phenomenal for a gage room), a replacement is found and the part is gaged. Now the parts being made are bad because the operator did not change the pressure to allow for the new roll of lapping paper. The pressure settings must be changed and the machine cleared of all nonstandard parts to make sure the quality is within standards. The changeover just lost another six minutes adjusting pressure, clearing out parts, and gaging quality. Sound like fun? It happens. How many manufacturing facilities can afford changeovers like this one? The line's current cycle time is 30 seconds; therefore, 126 parts of production were lost for a 15-minute changeover.

ADDRESSING THE ISSUES

Now that the current situation is known, the desired state should be targeted. Begin with the machine that stopped without warning. Sometimes this can be remedied by simply adding a photo-eye or proximity switch to a role of lapping paper to give the operator enough time to respond and prepare for the changeover (see Figure 14-2).

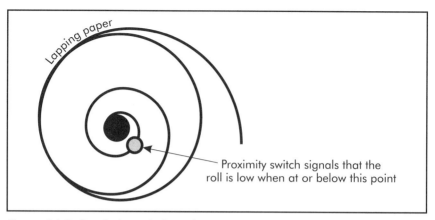

Figure 14-2. Proximity switch on lapping paper.

Next, address the time used to walk to the tool crib and wait for items. The operator spent 13 minutes (26 parts) getting the lapping paper needed for the tool change. Why not simply stage the lapping paper by the machine? This can be done with something as simple as a hook to hang the paper on, or it can be as efficient as a Kanban-type ordering system. Since each company has its own tooling/reordering system in place, using a current system would be advantageous to any company.

Address the rest of the waste in the example using one simple idea—why not make a tool change cart containing all the tools, gages, and materials needed, already color-coded and in the order they are needed? This can be done by developing a multifunctional cart commonizing set-up tooling, gages, and other materials. Re-evaluate the new procedure versus the old in Table 14-2.

Quantifying the Current State

Based on the preceding example, the time savings can be equated to dollars to easily understand the substantial savings possible. Assume that lapping is the final operation in the process. A lapped part costs $40, but can be sold to a customer for $75. Therefore, a $35 profit is made for every part sold. The previous example of changeover cost 126 parts:

63 minutes of downtime × 30 seconds/cycle = 126 parts

Table 14-2. Lapping paper change: old way versus new way

Old Way	Downtime (min)	New Way	Downtime (min)
Waiting on operator	5	Proximity switch to warn operator	0
Walk to crib	3	Paper at machine	0
Waiting in line	7	Paper at machine	0
Return to machine	3	Paper at machine	0
Locating tools	5	Tools on cart in order	0
Tool change	15	Assuming no internal improvement	15
Locating gage	4	Gage on cart	0
Waiting on new gage	15	Gage color-coded	0
Quality issue/pressure	6	Posted on cart/machine	1
Total	63		16

A profit margin of $35 per part figures out to be $4,410 based on selling 126 parts for one simple tool change. Annualized (assuming the facility works 220 days and performs the changeover once per day), it could cost a company $970,200. Naturally, not all opportunities are going to be this dramatic, but similar cases have been documented. Here are two simple questions to ask wherever changeover opportunities exist:

1. What can be prepared in advance to reduce downtime?
2. Would the improvement affect the internal setup?

Personnel Options

What about manpower? Are employees being used to the fullest? Perhaps two operators could be used at one time for the same changeover. Manpower can play a vital role in reducing changeover times. Adding more manpower or performing multiple changeover activities at once may be a logical approach to reducing machine downtime. For example, a process has a tool life of 6,250 pieces, while the preceding process produces only 5,750 pieces per tool. Depending on tool cost and tooling reliability, it may be more beneficial to use both tools at 6,000 pieces per tool, which will reduce the amount of machine downtime for changeovers.

Another example of efficient manpower is similar to the relationship of a surgeon and an assistant. One person may perform the actual changeover while the other makes sure that all the tools and parts are ready when needed. Any non-value-added movement to this process should be eliminated.

Standardizing Functions

One method of shifting from internal to external activities at changeover is to use functions that aid in standardization. Standardized applications may apply to elements such as securing, gaging, centering, gripping, or expelling. There are two basic steps to implementing standardizations for setup:

1. Look at each setup or function that is common to all processes.
2. Look at the setups and functions that may be commonized or made more efficient using the fewest number of tools or parts.

Naturally, the quickest way to perform changeovers is with the fewest parts changes (or no changes). An example of this application may be a part that has common machined dimensions, but with totally different raw dimensions for various clearances for different customers. Assume steering gears for two different car manufacturers are made. Each manufacturer uses the same basic parts with one exception: the valve body on one is several millimeters shorter for mounting purposes. During valve insertion, each product undergoes three basic steps: gripping the valve, inserting the valve, and returning to the original position. When changing to a new product, only the insertion jig needs to be changed to match the new dimension, shape, or material. There is no need to change the insertion travel or return stroke. The jig should be dimensioned to allow for the length difference.

Centering Dies

Other ways to move from internal to external activities include the use of jigs to center dies. When setting up a press, a die may need to be positioned in the center of the bolster. Without the use of a jig this can be time-consuming and costly to the operation. Remember, the longer a machine is down for changeover, the fewer parts it produces. Some dies have shanks on the top that fit exactly into the shank attachment hole in the ram of the press. Traditionally, to get the die centered for alignment, an operator might inch the ram downward while aligning the shank and hole with the naked eye. This process takes a lot of time and patience because any damage may render the die inoperable. The process can be improved by attaching a centering jig to the machine so it is a fixed distance from the center of the die and shank. The use of a V-type jig is most useful when trying to align the right and left adjustment to the center of the jig. When attaching a male V-type jig to the machine, the female portion also must be attached to the die. When the two are pressed together, assuming the V-jigs have been set up properly, a perfectly centered jig should be attained that aligns with the shank and holds up perfectly. Setting a die becomes a simple operation, drastically reducing errors and set-up time.

Intermediary Jigs

Intermediary jigs also may be used to change an internal setup to an external setup. An *intermediary jig* is defined as a plate or frame of standard dimension that can be removed from the machine. One of these jigs may be used to attach a die used in a machine. Meanwhile, another die can be set up externally with a second jig. When it is time for changeover, the die is already attached, centered, and ready for internal change. One example of this is in a large press that requires multiple dies. Two identical intermediary jig plates can be made to hold the dies. While the first set of jigs and dies is hard at work making product in the machine, two new dies are attached to the other intermediary jig and centered. When changeover occurs, the operator or jobsetter simply removes the old die and jigs and replaces them with the preset dies and jigs. This way, the internal steps can be completed with the machine down only for a short time.

STREAMLINING INTERNAL AND EXTERNAL ACTIVITIES

The process of streamlining merely involves looking at ways to improve processes before performing them. This includes anything that takes place before the machinery is stopped, while it is stopped, and after it is running again on the new setup. It could include:

- visual controls, such as tooling standardization and color-coded tooling to match the changeover;
- implementing quick-type disconnect tooling, or
- parallel operations.

Other streamlining techniques may include reducing securing devices, such as using six bolts to secure a die instead of eight. Doweling may be an option to ensure proper alignment. Any method that reduces mistakes or unnecessary motion will help streamline changeover processes. Even simple things, like moving needed materials physically closer to the process, reduces the amount of time it takes for the changeover.

What is the best way to fasten dies or tooling to the machine? Bolts are traditionally used. However, bolts are time-consuming and costly for changeovers. The process of loosening and fastening bolts tends to slow internal change times, which results in lost opportunities to make products. Bolts pose many other problems. They may become lost, worn out, or even stripped, which can create a critical situation when they ruin the threads inside a machine. Hours of production can be lost tapping or repairing a damaged bolt hole.

Functional Clamping Devices

To avoid wasted effort, functional clamping devices may be used. A *functional clamp* holds objects in place with little effort. It can be loosened or tightened quickly. In addition to being much faster, a functional clamp may also remain on the machine, reducing the chance of getting lost or replaced by the wrong part. (How many times has the wrong-size bolt been used after losing the original, only to find that the new bolt cannot function as the original did?)

Functional clamping methods include one-turn and one-motion processes.

One-turn Clamping

The first example of one-turn clamping involves a pear-shaped hole. How many times has a bolt been completely removed when a pear-shaped hole could have been formed to simply turn a bolt one or two turns to remove a cover, guard, or die? What about the U-slot method? The *U-slot* method consists of simply cutting a slot in the edge of a die. By inserting the head of the bolt into a dovetail groove on the machine bed, the bolt can be slid into the U-slot of the die. Once again, the die can be tightened with one or two turns of the nut. The clamp method can even be used to reduce tightening time. An L-shaped clamp can be added to attach the die or tooling by one or two turns of the clamp bolt.

One-motion Methods

Now that bolt-tightening has reduced changeover time, one-motion methods can be implemented. Cam clamps are an excellent tool for securing the workpiece to the machine. Parts can be effectively secured into position with the simple flip of a handle. Another idea is the use of a spring when needing a pin or a stop to move into place. Other methods may include the use of taper pins and wedges. Magnets and vacuum suction are also convenient as one-motion methods. With just the flick of a switch, two smooth flat surfaces can become inseparable.

Adjustments

Streamlining internal and external activities includes not only fastening, but adjustments as well. Too much time is often spent making adjustments while machines are down that could have been used making parts during an external setup. This is an area in which 40%–60% of changeover time has been spent trying dies and making adjustments. Companies need to think "out of the box" again. Instead of trying to reduce adjustments, they need to think in terms of eliminating adjustments.

Gaging

Before starting up the equipment, checks should be performed to make sure centering and dimensioning are absolutely a cut above the rest. Preventive maintenance must be performed on gages to keep them in tiptop shape by replacing or repairing worn parts, verifying gage accuracy, and standardizing methods. Gage the scales with the appropriate application. When trying to hold a part tolerance within 50 microns, do not use a 100-micron indicator.

When gaging is not applicable, use the scaling method. Scales are especially useful when settings do not need to be as accurate as those with a dial indicator. The rule of thumb is that a scale is usually accurate to within 0.02 in. (0.5 mm). Mount a scale directly on a machine/slide fixed position, locate the ideal position, and mark it by applying a permanent alignment mark on the die with the premarked scale fixed to the machine. There will be no question as to the die position. Dial indicators may be used when a more accurate setting is needed. Dial indicators are typically accurate to within 0.0004 in. (0.01 mm). For companies on a tight budget, gages can be a financial burden. But for those companies that can afford it, digital numerical devices may achieve greater precision than imagined.

CONCLUSION

Once companies have implemented the changes suggested for quick changeover, it is vital that they are standardized so all personnel are aware of and trained on the new processes. If a company does not have such documentation, it should be developed for the quick changeover process. A simple checklist can be used to document the steps of a changeover process. Good visuals, with a digital camera or Polaroid pictures, can be useful when trying to document procedures. Documenting the current state is not only a plus for today, but for tomorrow as well. When new operators come in, they must have some sort of reference for training and standardization. A standardized changeover process is not the end, but a beginning.

Total Productive Maintenance 15

By Charles Robinson

Total Productive Maintenance (TPM) is the main driver that establishes stability in the manufacturing process in most lean implementations. Practicing TPM means using plant-floor activities to continuously improve equipment and manufacturing process reliability. The infrastructure of the classic TPM implementation discussed in this chapter should be a part of the overall lean implementation. Use the essence of TPM, as supported by the infrastructure of lean, to help ensure the success of the lean implementation.

DEFINITION OF TOTAL PRODUCTIVE MAINTENANCE

TPM has been in use in Japan since 1971, and was first introduced to North America in 1988. TPM is described as "productive maintenance carried out by all employees through small group activities." The theory is that maintenance should be shared by operations as well as the maintenance department. Just as a person is responsible for his or her own health, so is an operator responsible for the health of his or her machine. The maintenance technician can be compared to a doctor. When the operator cannot effectively treat the symptoms, a special expert must be brought in (either a doctor or a maintenance technician) to effectively diagnose and treat the abnormality (Nakajima 1988).

TPM is also defined as follows: "TPM is a plant improvement methodology that enables continuous and rapid improvement through use of employee involvement, employee empowerment, and closed-loop measurement of results (Robinson and Ginder 1995).

The classic TPM methodology, created by Nakajima, provides 12 steps for improvement in a cookbook method designed to take an organization from average to full TPM implementation. Supporting checklists, standards, and worksheets are used where appropriate. The methodology is structured, requires discipline, and has specific standards to judge performance. Hundreds of companies have adopted TPM as a strategic initiative to improve performance. The power of TPM has saved companies billions of dollars during the last decade.

Lean implementation starts with current state mapping and business case development, described in Chapters 4 and 5, respectively. After a plan is developed, the organization undertakes a five-phase implementation process (described in Chapter 7). The first phase achieves stability—it requires consistency in processes and machinery. TPM is the major driver for achieving machine stability. Without some level of machine stability, it is impossible to advance to the next phases of implementation without incurring repeated and painful returns to the stability phase.

In simple terms, the inventory, whether it is work-in-process, raw materials, or finished goods, is in the system for a reason. Usually it is built into the system as an insurance policy against instability or uncertainty in the system. If an operator is supplied by a machine with unpredictable reliability, it is only logical that he or she will insist on more inventory to compensate for that instability. If the supplying process is predictable and reliable, there is no need for an inventory buffer. Equipment reliability is a necessary prerequisite to obtaining stability. The best method, or tool, for obtaining equipment reliability is sustained TPM.

TRADITIONAL TPM STEPS

TPM methodology was developed by Nakajima in his work for a first-tier supplier. He designed TPM to be a stand-alone process. It was not meant to rely on any other tool or process to be successful. The traditional four-stage, 12-step implementation of TPM is shown in Table 15-1.

Table 15-1. The 12 steps of TPM

Stage	Step
Preparatory	1. Announce top management's decision to introduce TPM 2. Launch an educational campaign to introduce TPM 3. Create an organizational structure to promote TPM 4. Establish basic policies and goals of TPM 5. Form a master plan for implementing TPM
Preliminary implementation	6. Kick-off TPM

TPM implementation	7. Improve the effectiveness of each critical piece of equipment
	8. Set up and implement autonomous maintenance
	   I. Initial cleaning
	   II. Preventive cleaning measures
	   III. Establish cleaning and lubrication standards
	   IV. General inspection
	   V. Autonomous inspection
	   VI. Process discipline
	   VII. Independent autonomous maintenance
	9. Establish a planned maintenance system
	10. Provide training to improve operator and maintenance skills
	11. Develop an early equipment management program

Stabilization	12. Perfect TPM implementation and raise TPM levels

Preparatory Phase

The preparatory phase of TPM implementation establishes the commitment to deploy the TPM process. Like all improvement processes, TPM cannot be successful unless it has the necessary management commitment and corresponding support. If management is not willing to commit the required resources and measure the performance of the plant with TPM, it will not be a sustainable improvement process. TPM recognizes that fact and devotes the first five of its 12 steps to the preparatory phase. This ensures a high level of management commitment.

Step 1—Announce Top Management's Decision to Introduce TPM

The first step is a defining moment for managers. They make a formal commitment to the plant, their division, the company's senior management, and the rest of the world that they are about to embark on this new and different process—Total Productive Maintenance.

Step 2—Launch an Educational Campaign to Introduce TPM

The entire plant needs a base level of knowledge about TPM, including management's commitment to the process. This is best accomplished through an educational campaign using the learn, use, teach, instruct (LUTI) method. Top management must learn the process and teach it to those who report to them. Those employees in turn teach it to the next level of the plant until all have a base level of understanding.

Step 3—Create an Organizational Structure to Promote TPM

Once the announcement has been made and a baseline of knowledge is established, resources must be dedicated and a structure organized to guide the plant through the TPM process. The organization should include plant champions (process leaders), coordinators, trainers, mentors, and process owners who have defined roles and responsibilities for the implementation.

Step 4—Establish Basic Policies and Goals of TPM

A *policy* is a rule or guideline for all to follow when making decisions. A *goal* is an expected level of measurable performance as a result of TPM implementation. One example of a policy is: "no one will lose their job as a result of TPM implementation." An example of a goal is: "we expect to get 25% more product from existing equipment with no increased overtime as a result of implementing TPM." Establishing these policies and goals is an important step in creating the organizational structure.

Step 5—Form a Master Plan for Implementing TPM

Step 5 is basic project management. A master plan establishes what is to be done, where, at what time, how long it will take, and the expected results. When developing the plan, managers must recognize that things will change and not necessarily go according to plan. In spite of this recognition, failing to develop a meaningful plan is a recipe for failure. Time should be spent to think through all of the requirements; a reasonable plan for accomplishment can be developed based upon the current level of knowledge. As the knowledge changes, the plan should be altered to reflect that new level of understanding. Step 5 completes the preparatory phase of TPM implementation.

Preliminary Implementation

Step 6—Kick-off of TPM

Why is the kick-off important? Perhaps it is better to define this phase as the celebration of TPM. Remember that TPM is considered a process that has a beginning but not an end—it is always improving. The kick-off is meant to be a celebration of the beginning of something new and good for the plant. It is a new way of doing business with regard to how equipment and processes are maintained. It cannot be celebrated at the end because there is no end.

TPM Implementation

The following six steps are designed to move the TPM from the planning process to the plant floor.

Step 7—Improve the Effectiveness of Each Critical Piece of Equipment

There are many pieces of equipment in any manufacturing process. Some are critical because they constrain processes or provide safety, quality, or reliability. Some are more critical than others. The basis of this step is that critical items need to have their effectiveness improved.

Effectiveness is considered in three ways:

1. Availability—can the equipment run when the operator wants it to run?
2. Efficiency—is the equipment performing at the rate for which it was designed?
3. Quality rate—is the equipment producing quality products?

The outcome of these considerations is termed overall equipment effectiveness (OEE), which is discussed in detail in Chapter 3. Improving equipment effectiveness is accomplished by examining and eliminating losses and wastes.

Step 8—Set Up and Implement Autonomous Maintenance

Autonomous maintenance encourages operator ownership of the equipment reliability function. Managers can encourage this by using a seven-level deployment scheme that slowly establishes and maintains "better than new" equipment condition. The seven levels of autonomous maintenance are critical in the TPM process. They establish the stability of equipment as a bedrock on which to build a lean implementation.

The seven levels of autonomous maintenance include:

1. *Initial cleaning* is required by all personnel associated with a piece of equipment. Beginning with where the product touches the machine and working outward, operators must clean surfaces with the objective of identifying and correcting defects. This is similar to the 5-S process in a visual factory system except that it focuses on machines as opposed to an area. Much like washing and waxing a car is the best way to locate all paint and body defects, cleaning equipment is the best way to spot machine defects. See Table 15-2 for an

Table 15-2. Level 1: initial cleaning performance checklist

Initials	Activity
	Perform all activities necessary to shut down, isolate, and make the equipment or process area totally safe.
	Obtain copies of equipment drawings, documentation, histories, and other relevant information. (This documentation is used in successive levels of maintenance to develop standards for lubrication and inspections.)
	Document initial condition of the equipment through photographs. Prepare forms for documenting equipment problems and tags for marking items needing further inspection.
	Segment portions of the machine or process area, and plan how to clean the machine or area with maximum efficiency and effectiveness.
	Obtain hand tools, rags, brushes, solvents, mops, brooms, scrapers, and any other tools required to perform the cleaning tasks.
	Clean each machine segment in a methodical manner. Return the equipment to an "as new" condition.
	Remove all dirt, grime, dust, grease, oil, sludge, chips, trash, and excess materials. Note any equipment abnormalities. Cleaning is inspecting!
	Tag and document all equipment abnormalities.
	Retorque all bolts, including hold-down, fastener, adjustment, and structural bolts. Mark all bolts by painting a stripe across both the stud head and the bolt to indicate their relative positions when properly torqued.
	Repaint areas if necessary according to predetermined repaint specifications. Color-code piping, utilities, and guards for ease of observation.
	Note and mark all lubrication points. Again, use color coding.
	Photograph the clean machine or process area and compare with a "before" photograph to verify progress.
	Formally turn the equipment or process area back to production for startup and operation.

example of an initial cleaning checklist. Table 15-3 is a sample of an initial cleaning work sheet.

2. *Preventive cleaning measures* help keep machines clean. This level consists of finding ways to control internal and external contamination. Internal contamination includes process excesses, such as chips from a grinding machine, and lubrication and process leaks. External contamination includes dust buildup, environmental debris, and human-generated trash. Controlling this contamination using preventive methods and periodic restoration makes equipment defects easier to spot. This increases the probability of safe operation and minimizes the potential for product contamination. See Tables 15-4, 15-5, and 15-6 for checklists, standards, and work sheets to complete this level.

3. *Develop cleaning and lubrication standards* once the machines are cleaned and maintained. This level establishes standards for cleanliness. Standards form the basis for control so that deviations are easy to spot and act upon. Standards can be pictures, checklists, or other visual factory methods. The standard should be posted on or near the machine and used on a regular basis. In addition to cleanliness standards, lubrication standards should be established. The lubrication standard forms the basis for all machine lubrication points, including specifications and frequency of checks and changes. Tables 15-7, 15-8, and 15-9 provide checklists, standards, and work sheet documents that further describe this level of autonomous maintenance.

4. *General inspection* should be conducted, reviewed, verified, and categorized once lubrication and cleaning standards are implemented. This identifies the appropriate actions and timing necessary to support the needs of the equipment and processes. General inspection should be performed by the operator, skilled maintenance tradespeople, and engineering, with some input from safety personnel. This establishes the real inspection needs of the equipment, identifies specific roles and responsibilities for personnel who are responsible, and creates a schedule for accomplishment. Table 15-10 is a general inspection checklist for this level of autonomous maintenance.

Table 15-3. Initial cleaning work sheet

Group Checked: _____ Date: _____
Checked by: _____

Items to Check	Possible Points	Actual Points	Notes or Remarks
1. Cleaning of line and machines			
1. Any dirt, dust, excess oil, damage, foreign objects?	5		
a. Moving parts, rotating parts			
b. Braking apparatus, locking mechanisms			
c. Guides, fixtures			
d. Frame, surfaces			
2. Any nuts, bolts, etc., loose, wobbly, missing?	10		
3. Any play in moving parts, mounting sections?	10		
4. Any unneeded objects on body of machine?	3		
5. Is the machine firmly seated?	3		
2. Cleaning of auxiliary equipment			
1. Any dirt, dust, excess oil, damage, foreign objects?	5		
a. Cylinders, solenoid valves, air			
b. Limit switches			
c. Motors, gears, shafts, couplings			
d. Piping			
2. Any nuts, bolts, etc., loose, wobbly, missing?	10		
3. Any leaks of oil, water, air, gas, steam, product?	5		
4. Any indicator lamps not working?	3		

Table 15-3. (continued)

Items to Check	Possible Points	Actual Points	Notes or Remarks
3. Lubrication			
1. Does oil reach all appropriate moving parts?	10		
2. Is the right amount of oil in reservoirs, etc.?	5		
3. Do oil apertures have covers?	3		
4. Are oil supply pipes clean, no leaks or seepage?	5		
4. The 5 Ss around the line and equipment			
1. Stabilize—orderliness (discard what you don't need)	3		
2. Sort—neatness (needed items are accessible)	3		
3. Shine—cleanliness (well-organized, clean)	3		
4. Standardize—adherance (and inspection)	3		
5. Sustain—discipline (following proper procedures)	3		
5. Status of activities			
1. Are Kaizen examples, goals, and results posted on the group's bulletin board?	3		
2. Do the corrective action plans address root causes and cleaning problems?	3		
3. Are daily checks properly performed?	2		
Total points possible	100		**Actual points earned**
6. Safety			
1. Any damage to safety devices, enclosures, shields?	yes*/no		
2. Any other unsafe conditions?	yes*/no		

*A "yes" answer to any safety item results in overall failure.

Table 15-4. Level 2: preventive cleaning measures
—process leaks checklist

Initials	Activity
	Locate the leak and identify its source, the type of fluid leaking, and the severity of the leak.
	Fill out the leak tag form, noting the information as well as your name and date.
	Detach the two-part form and affix one portion to the leak or in near proximity of the leak.
	Send or hand the other part of the form to the group responsible for backlogging work orders.
	Document the abnormality of the machine/equipment problem listing, and record the tag information on the tag log.

5. *Autonomous inspection* includes all the inspection activities that can be effectively and safely performed by operators with appropriate training. This level institutionalizes the operator's performance of inspection activities. It is important to note that although skilled maintenance tradespeople are relieved of their responsibilities in this matter, it does not mean that their role will decrease in the plant. Their duties transition into a more technical role of performing precision repairs and focusing on improving equipment effectiveness through analysis and corrective or improvement activities. In addition, skilled tradespeople can be required to help specify and upgrade new equipment installations to ensure that they are maintainable and reliable. Table 15-11 is a checklist for accomplishment of the autonomous inspection level.
6. *Process discipline* must be established once the preceding processes have been defined and implemented. Processes need to be institutionalized to become independent of most variables. They must be the accepted method of operating the business, regardless of who the plant manager is or what product changes are forthcoming in the plant. This process becomes accepted as the standard. Machinery is cleaned and

Table 15-5. Level 2: standards for problem causes and inaccessible areas

Item No.	Point Scale			Evaluation Guidelines		
	L	M	H	Low	Medium	High
Sustaining the results of the initial cleaning						
1.	1	2	3	Main pieces of equipment show dirt, wobble, unneeded items	Dirt and wobble evident when touched Unneeded items in good order	No dirt or wobble No unneeded items
2.	1	2	3	Auxiliary equipment shows dirt, wobbles, etc.	Dirt and wobble evident to touch	No missing, loose, or wobbly parts
3.	1	2	3	Oil inadequate, not circulating completely	Partial oil gaps	Adequate oiling and circulation
4.	0		1	Very poor housekeeping (5 Ss)		Good housekeeping (5 Ss)
Status of corrective action to eliminate causes of dirt, leaks, dust, etc.						
1.	1-2	3	4-5	No charts	Partial/outdated charts	Up-to-date, well-utilized charts
2.	1-2	3	4-5	Action at 60% complete	Between 60% and 80% complete	More than 80% complete
3.	1-2	3	4-5	Many leaks	Leaks present but contained	No leaks
4.	1-2	3	4-5	No plans	"Intention" but no organization	Plans organized
5.	1-2	3	4-5	Two or more unknown causes	One unknown cause	No unknown causes

Table 15-5. (continued)

Item No.	Point Scale L	Point Scale M	Point Scale H	Evaluation Guidelines Low	Evaluation Guidelines Medium	Evaluation Guidelines High
Improvement of accessibility						
1.	1-2	3	4-5	No charts	Partial/outdated charts	Up-to-date, well-utilized charts
2a.	1	2-3	4	No special tools devised, though needed	Special tools being designed	Appropriate special tools used
b.	1	2	3	No action to make cover removal easier	Some action being taken	Special improvements implemented
c.	1	2	3	No progress; housekeeping still difficult	Some action to streamline housekeeping	Housekeeping has been made easy
3.	1-2	3	4-5	No plans	"Intention" but no organization	Plans organized
4.	1-2	3	4-5	Two or more inaccessible areas remaining	One inaccessible area remaining	No inaccessible areas

Table 15-5. (continued)

Item No.	Point Scale			Evaluation Guidelines		
	L	M	H	Low	Medium	High
Management						
1.	1-3	4-6	7-10	Individual machine goals, no overall goals	Some overall goals, but insufficient detail	All individual goals linked to overall goals
2.	1-2	3-6	7-10	In preparatory stages	Written, but lacking sufficient detail	Sufficiently detailed and complete
3	1-2	3-6	4-5	Checks performed but results not posted	Results posted but checks are irregular	Checks according to plan and results posted
4.	1-3	4-6	7-10	No evidence of active use	Activity posted but out-of-date	Bulletin boards used and up-to-date
Follow-up on actions from initial cleaning						
1.			10	Action less than 70% complete	Action more than 70% complete Remaining 30% to be completed within one month	Remaining 10% to be completed within six months

Total Productive Maintenance

Table 15-6. Level 2: follow-up for problem causes and inaccessible areas work sheet

Group Checked: _____ Date: _____
Checked by: _____

Factors to be Checked	Possible Points	Actual Points	Remarks
Sustaining the results of the initial cleaning			
1. Is the equipment still in the state achieved by the initial cleaning?	10		
Status of corrective action to eliminate causes of dirt, leaks, dust, etc.			
1. Are causes of dirt and other problems summarized on charts?	5		
2. Has corrective action been taken to address the causes of dirt, dust, leaks, and other problems?	5		
3. Has action been taken to correct leaks?	5		
4. Are there plans to deal with the remaining problems?	5		
5. Are there any remaining problem causes that haven't been indicated?	5		
Improvement of accessibility			
1. Are inaccessible areas summarized on charts and are there regular progress checks?	5		
2. Have areas been improved to make cleaning easier?	10		
a. Have special cleaning tools been devised?	(4)	( )	
b. Has cover removal and replacement been made easier?	(3)	( )	
c. Does housekeeping concentrate on the proper activities?	(3)	( )	
3. Are there plans to deal with the remaining problems?	5		
4. Do other inaccessible areas remain?	5		

Table 15-6. (continued)

Factors to be Checked	Possible Points	Actual Points	Remarks
Management			
1. Have goal times been set for cleaning and lubricating according to the established cleaning and oiling standards?	10		
2. Have written cleaning and oiling standards been completed?	5		
3. Are daily checks and follow-ups being performed?	5		
4. Are bulletin boards in active use for posting improvement examples, activities, goals, results, etc.?	10		
Follow-up on initial cleaning	10		
Total points possible	**100**		

Table 15-7. Level 3: development of cleaning and lubrication standards checklist

Initials	Activity
	Convene small groups under the guidance of a facilitator and create a vision of the equipment's condition as "better than new." Gather and review all relevant materials, such as manufacturers' manuals, drawings, existing preventive maintenance check sheets, or historical operational data.
	Describe what each component of the equipment would look like if it were in "better than new" condition.
	Use existing drawings or develop new ones that visually show the equipment. Label each component of the equipment and identify each lubrication point on the picture.
	For each component, document the "better than new" condition using terms such as: free of all oil, painted, free of all dust buildup, and free of process excess.
	Establish and document the frequency of cleaning or inspection required to maintain this "better than new" condition.
	For each lubrication point, document the following: • Lubricant required • Reservoir capacity • Filter requirements • Frequency of checks • Frequency of sampling • Frequency of change • Safety considerations

* All machine lubrication points should be physically marked. Use color codes corresponding to the type of lubricant to be used at that lubrication point.

Table 15-8. Lubrication standards work sheet

Item No.	Point Scale			Evaluation Guidelines		
	L	M	H	Low	Medium	High
Follow-up						
1.	1-2	3	4-5	Uncorrected malfunctions persist	Malfunctions identified and corrective actions being deliberated	Adequate follow-up and successive improvements
2.	1-2	3	4-5	Not carried out according to standards	Carried out according to standards but equipment not kept up adequately	Adequate follow-up and accumulated improvements
Status of group activities						
1.	1-2	3	4-5	Fewer than 90% have received training	More than 90% have received training	100% have received and understood training
2.	1-2	3	4-5	Some take part; participation weak	Participation OK; attendance poor	100% attendance and participation
3.	1-2	3	4-5	Maintenance personnel do not participate	Some checks made with maintenance personnel	Adequate checks made with maintenance personnel

Total Productive Maintenance

Table 15-8. (continued)

Item No.	Point Scale L	Point Scale M	Point Scale H	Evaluation Guidelines Low	Evaluation Guidelines Medium	Evaluation Guidelines High
Status of group activities						
4.	1-2	3	4-5	No priorities set for discovered malfunctions	Priorities set but improvements slow	Improvements proceeding according to plan
5.	1-2	3	4-5	No check sheets developed or inspections made	Inadequate check sheets and inspections	Check sheet developed and inspections made
6.	1-2	3	4-5	Not added to standards; cannot be adhered to	Some revision of standards needed	Standards suitable and adequately adhered to
Lubrication standards						
1.	1-2	3	4-5	Hard to see	Somewhat hard to see	Easy to see
2.	1-2	3	4-5	Oil leaks, dirt present; control limits not set	Improvement not yet adequate	Oil volumes and temperatures controlled; no leaks or dirt
3.	1-2	3	4-5	More than two items off or missing	One item off or missing	100% OK
4.	1-2	3	4-5	Oil and grease leaks visible	Oil, grease visibly oozing	100% OK

343

Table 15-8. (continued)

Item No.	Point Scale			Evaluation Guidelines		
	L	M	H	Low	Medium	High
Lubrication standards						
5.	1-2	3	4-5	Two or more parts with oil	One part with oil	100% OK
6.	1-2	3	4-5	Two or more parts with oil	One part with oil	100% OK
7.	1-2	3	4-5	Conspicuous filth due to excessive oil; poor cleanup	More improvement needed	Suitable oil supply and good disposal practices
8.	1-2	3	4-5	Less than 90% specified lubricant	Specified lubricant used 90% or more	Lubricants used as specified
9.	1-2	3	4-5	Amounts unknown; schedule unchanged	Amounts known, but improvement needed	Amounts known, anti-leak steps, oiling schedules
Follow-up actions from previous lube check						
1.	1-5	6-10	11-15	Action less than 70% complete	Action more than 70% complete Remaining to be completed in one month	Action more than 90% complete Remaining to be completed within six months

Table 15-9. Follow-up for lubrication standards work sheet

Group Checked: _____ Date: _____
Checked by: _____

Factors to be Checked	Possible Points	Actual Points	Remarks
Follow-up			
1. Are improvements made so far being maintained?	5		
2. Are cleaning, oiling, and maintenance tasks being carried out as specified, and is equipment being kept up?	5		
Status of group activities			
1. Is education proceeding according to plan?	5		
2. Are meeting frequencies, attendance rates, times OK?	5		
3. Were inspections made using maximum personnel and check sheets?	5		
4. Have malfunctions been corrected?	5		
5. Are check sheets developed and inspections made as a group?	5		
6. Are items to inspect added to written standards; are they suitable in terms of content, frequency, and distribution, and can they be adhered to? (Are they being adhered to?)	5		

Table 15-9. (continued)

Factors to be Checked	Possible Points	Actual Points	Remarks
Lubrication			
1. Are oilers clean inside and out? Is the oil level easily visible?	5		
2. Are oil quantities and temperatures suitable for oil tanks, etc.? Any oil leaks, dirt?	5		
3. Any sites with damaged or missing grease nipples, oil caps, etc.?	5		
4. No oil or grease leaks from supply devices, pipes?	5		
5. Is there oil on rotating, sliding, and drive; for example, chain parts?	5		
6. Is there lubricant on air cylinder rods?	5		
7. Is excess oil fouling the equipment?	5		
8. Is the specified lubricant being used?	5		
9. Is it known how much lubricant is being used, and is that amount being reduced through reduction of leaks?	5		
Follow-up on actions from previous check	15		
Total points possible	100		Actual points earned

Table 15-10. Level 4: general inspection checklist

Initials	Activity
	Collect all available equipment data, including preventive and predictive maintenance procedures, historical operational data, historical failure data, vendor/manufacturers' recommendations, and all design drawings/specifications.
	Review existing preventive maintenance, predictive maintenance, and inspection procedures for accuracy and completeness.
	Review all equipment failures and isolate their root causes.
	Review manufacturers' and vendors' recommended checks and maintenance procedures.
	Document the general inspection standards in a format similar to that used for the lubrication and cleaning standards.
	Categorize all inspection procedures into those that can be performed presently by production operators, those that can be performed by production operators with training, and those that should be performed by craftspeople.
	Categorize inspections by: • the function or craftsperson performing the inspection; • whether the equipment must be operational or shut; down while the inspection is being performed; • whether disassembly is required; and • ease of performance.
	Develop check sheets for the above inspections.
	Identify all training required to educate production operators on the safe performance of designated tasks.

Table 15-11. Level 5: autonomous inspection checklist

Initials	Activity
	Develop checklists for each operator position and organize them by performance interval (for example, daily, weekly, monthly).
	Distribute to various personnel and/or shifts so that the workload is spread evenly across workers and variances in operating conditions.
	Use pictorials to illustrate the tasks to be performed.
	The maintenance management system should be used to schedule checks that occur at intervals of one week or greater. This will provide an electronic record of the check and related performance.

lubricated according to standards, and inspections are made according to the standards with accepted responses to abnormalities. This area is given its own level mainly because it takes time and continued support to make any new practice or procedure the accepted standard.

7. *Independent autonomous maintenance* continuously improves the maintenance process by searching for and eliminating waste from within the maintenance process. In essence it improves the standard developed and established in the previous six levels of autonomous maintenance. Table 15-12 shows a checklist of activities.

Step 9—Establish a Planned Maintenance System

Most plants may claim to have a planned system of maintenance. Unfortunately, the system is often limited to activities scheduled in a preventive maintenance system. As in most areas of lean manufacturing, whichever system is established should focus on the elimination of waste. Studies have shown that planned work takes less than 25% of the time and effort as unplanned work. Therefore, to eliminate waste, plan as much maintenance work as possible. Planned maintenance work enables the early detection of poten-

Table 15-12. Level 7:
independent autonomous maintenance checklist

Initials	Activity
	List potential work processes that are perceived to be less than optimal in terms of effort or resource.
	Map the present method of performing work in terms of activities, decisions, and interactions with equipment, people, and other systems.
	Examine the work-flow diagram to identify multiple ways of accomplishing the same task.
	Develop the most effective procedure by eliminating redundant, non-value-added steps, and reducing time and resource requirements.
	Modify current, documented, work-flow diagrams to reflect upgraded methods of performing the work.
	Verify that the upgraded method is superior to historical methods by testing it on the plant floor.
	Train all appropriate workers in the new methods.
	Repeat the entire process as required until all work procedures in the plant have been evaluated.

tial equipment problems so corrective actions can be taken before they surface. This requires a dedicated resource to perform a number of functions.

- Analyze all collected equipment data for potential problems. This includes feedback from skilled tradespeople performing preventive maintenance, operators performing inspections, and special data from predictive techniques, such as vibration analysis, lube oil analysis, and infrared surveys.
- Determine potential problems and appropriate corrective actions.
- Determine resources, tools, data, and parts required to perform the corrective action.

- Determine a reasonable schedule for performing the corrective action, based upon production requirements, and equipment and safety risk.
- Determine an appropriate schedule with input from the production and maintenance personnel.
- Accumulate and stage all resources to perform the job in accordance with the schedule.
- Monitor the job performance.
- Collect historical data for future reference.

Anything less than the preceding functions minimizes the benefits of maintenance planning.

Step 10—Provide Training to Improve Operator and Maintenance Skills

As with most improvement methods, TPM focuses on training; specifically, it spotlights the upgrading of skills for both operators and tradespeople. The objective is for operators and skilled tradespeople to transfer their skills to each other. This is not meant to eliminate classifications, but to provide a base level of knowledge from which equipment performance can be continually improved.

Step 11—Develop an Early Equipment Management Program

Most aspects of equipment reliability are determined during the equipment specification and design phase of the equipment life cycle. TPM promotes several measures and methods for increasing the reliability of equipment and processes. It does this through focusing on reliability and up-front involvement by skilled tradespeople early in the equipment procurement process.

The engineering department usually has a budget goal for any new equipment installation that encourages the lowest total installed cost. The production department inherits that lowest total cost, and has the goal of producing to the customer demand. The goal of maintenance personnel is to keep the equipment running (at whatever cost) to meet the needs of production. These goals may not be totally aligned. Early equipment management specifies the ultimate aligned goal of lowest total equipment life cycle cost. Minimizing the total life cycle cost is clearly the best business decision for the organization.

Other metrics that measure the reliability and maintainability of the equipment include mean-time-between-failure (MTBF) and mean-time-to-repair (MTTR). MTBF measures the rate of equipment failure—how reliable is the equipment? MTTR measures how long repairs typically take—how maintainable is the equipment? Early equipment management stresses that these questions be asked before the equipment purchase is made. They should be integral selection criteria for any equipment purchases.

Early equipment management also promotes the use of existing knowledge bases of equipment performance to assist in developing specifications for future equipment purchases. Skilled tradespeople should provide data to knowledge bases based upon their experience in operating and maintaining the equipment. This knowledge must be documented to ensure that the next generation of machine purchases does not replicate earlier mistakes. Documentation requires much more than asking others for their opinion. It is a structured method of using the experience base to make intelligent specification and purchase decisions.

Stabilization Phase

TPM is a process, not a program. It is not something to start and stop once the steps have been completed. It has a beginning but no end. Companies that have treated TPM as a program get benefits for awhile, but quickly revert back to old practices and behaviors once the program ends. The stabilization phase is what makes TPM a process and not a program.

Step 12—Perfect TPM Implementation and Raise TPM Levels

Step 12 begins the continuous improvement portion of the process. Continuously repeating the steps is the best method of continual improvement. Challenging the workforce to set its own continually higher goals for achievement will drive the process to higher levels.

The preceding 12-step process has been successfully implemented by a number of companies across the world. The Japan Institute of Plant Maintenance (JIPM) recognizes company achievements in TPM by issuing awards for successfully practicing the process. JIPM has even established a branch office in Atlanta to promote TPM in the United States. In the spirit of continuous improvement, the JIPM has created three levels of awards for plants to achieve. The awards are highly coveted in Japan and provide focus for many plants to continuously improve. Non-Japanese plants also have been awarded the JIPM award for their TPM achievements. In fact, the highest TPM award has only been achieved by one plant—the Volvo assembly plant in Belgium.

MOLD THE TPM EFFORT INTO A LEAN IMPLEMENTATION

The standard 12-step TPM process was developed as a standalone process. Its details were not designed to fit into a lean implementation approach. Many of the steps developed to enable TPM to stand on its own are not required if TPM is part of an overall lean effort. Mold all activities designed to develop infrastructure for the TPM effort into a lean plan's overall infrastructure development. All activities that are part of improving the plant floor should be directly inserted as well. The following list contains the standard 12-step TPM process with suggestions on how to modify the steps to make them a part of the overall lean implementation effort.

1. *Announce top management's decision to introduce TPM.* This should be made a part of the overall lean effort. Management should be supportive of TPM as one of the tools of lean.
2. *Launch an educational campaign to introduce TPM.* Include this as part of the overall educational campaign for lean. Introductory courses for the plant floor workers should focus on lean. Position TPM as one of the first tools deployed because equipment must be stable before significant improvements can be made.
3. *Create an organizational structure to promote TPM.* The organizational structure should support lean without a particular focus on TPM. A part of the lean structure can be devoted to TPM, but it is a piece of the puzzle, not the entire solution.

4. *Establish basic policies and goals of TPM.* Policies and goals should focus on the entire system. Some policies and goals focus on TPM but they should be viewed as a part of the whole.
5. *Form a master plan for implementing TPM.* The master plan should be for implementing lean. The TPM plan should support the master plan.
6. *Kick-off TPM.* The process is lean so the kick-off should be for lean. The first six steps of TPM address infrastructure development. Implementers should focus on infrastructure development for a lean plant, and have the lean infrastructure support TPM implementation where appropriate.
7. *Improve the effectiveness of each critical piece of equipment.* This is an important step in achieving stability for the manufacturing process. To best support lean, "critical" equipment should be defined as being capable of constraining the manufacturing process. There are two types of constraints—physical and operational. A *physical constraint* is an operation with the ability to limit the output of the manufacturing process (the operation with the lowest stated capacity in the value stream). An *operational constraint* limits the output of the process due to inefficiency (it would not be a constraint if it were operated at the stated capacity). These constraints may be different in a single manufacturing line. For example, if there are three machines in sequence with respective capacities of 80, 60, and 100 parts-per-minute (PPM), the second operation (60 PPM) is the physical constraint. If the respective efficiencies of the machines were 50%, 90%, and 75%, the operational constraint would be the first machine (50%) since it has the lowest efficiency times capacity. It would limit the line capability to 40 parts-per-minute (80 parts-per-minute at 50% efficiency). In this case, the critical machine that needs its efficiency improved is the first machine, which is not the physical constraint.
8. *Set up and implement autonomous maintenance.* The seven steps of autonomous maintenance discussed earlier are the heart of TPM plant floor activities. It is extremely important to the lean effort. The first three steps should be accomplished during the stability phase. The remaining four

steps can follow during the rest of the five-phase lean implementation process.

9. *Establish a planned maintenance system.* A planned maintenance system greatly enhances system stability. This step should begin in the stability phase and continue as part of the overall process. Most plants have some form of planned maintenance system. They must continually examine their system for effectiveness and validity.

10. *Provide training to improve operator and maintenance skills.* The training described in this step should be included in the overall lean training curriculum. It should begin during the continuous flow phase and continue throughout the process.

11. *Develop an early equipment management program.* Reliability and maintainability must always be considered in any equipment specification and purchase. These requirements should be incorporated as part of the overall design methods for lean. (See Chapter 18 on greenfield implementation for lean.)

12. *Perfect TPM implementation and raise TPM levels.* Concentrate on continuous improvement of lean as the relentless pursuit of elimination of waste from the manufacturing process.

MAINTENANCE PROCEDURES AND STANDARDIZED WORK

The TPM process defines standard maintenance practices to maintain equipment. The procedures must be followed by both operators and maintenance's skilled tradespeople. There are procedures for equipment and lubrication checks, replacement of worn parts, inspections, and cleaning.

Lean manufacturing has a systematic method for capturing the best methods of performing various activities, which is called standardized work (described in Chapter 13). Most maintenance departments try to develop their own forms and formats to capture the steps of these inspections. There are even some elaborate computerized maintenance management systems that have established formats for inspections. A better method of defining these standard requirements is through the use of periodic work activities forms—a form of standardized work. This provides a method of not only standardizing inspections, but the methods used by operations and maintenance.

CONCLUSION

TPM is an integral part of all lean implementations. It is a valuable tool used to develop and maintain equipment and process reliability. Predictable equipment and process performance is a prerequisite for advanced implementations of lean. TPM enables low work-in-process inventory levels, higher product quality, and reduced time from order to delivery. TPM uses empowered plant floor personnel acting together to control and improve their workplace. TPM as a standalone process is powerful. TPM, working as a component of lean implementation, is world class.

REFERENCES

Nakajima, Seiichi. 1988. *Introduction to TPM: Total Productive Maintenance*. Portland, OR: Productivity Press.

Robinson, C.J. and Ginder, A.P. 1995. *TPM: The North American Experience*. Portland, OR: Productivity Press.

Problem-solving   16

By Robert Mussman

In today's lean manufacturing facility, one of the main responsibilities of both line operators and managers is to recognize, react to, find a solution for, and learn from day-to-day problems generated throughout their areas. The main concept of problem-solving is to "put the problem to bed." In other words, to solve the problem so it never returns. In many organizations, problems are viewed as opportunities for improvement, not as crises. This chapter walks problem-solvers, whether they are managers, supervisors, or team leaders, through the problem-solving process. Later in the chapter the process is applied in an extended illustration of the plan-do-check-adjust (PDCA) process.

THE PROBLEM-SOLVING MENTALITY

The problem-solver must determine what the specific problems are before he or she can begin to solve them. A "consciousness" of the problem must be developed; a process whereby the problem-solver determines a problem prior to it becoming a critical, plant-stopping, customer-losing issue. The problem-solver must understand *triggers*, advance warnings that things are not as they should be. The following can be used as triggers of problems with plant safety:

- Are injury data reports reviewed?
- Is staff compliance with safety regulations, including safety gear, checked?
- Are processes observed to ensure they are safely being performed as prescribed?

Triggers of quality problems are found by checking the following:

- Are defect reports reviewed with an eye toward finding repeating items?
- Are the processes checked? Look especially at goods produced at start-up, after lunch, and at breaks.
- Are the completed products or processes checked on a regular basis?

Triggers of production quantity problems may come from the following:

- Is production checked periodically? How often?
- If an Andon system is available, are the processes that cause the most downtime observed?
- How is the productivity rate? Is it constant or does it fluctuate?
- Can process steps be regrouped for an even distribution of work?
- Have scrap reports been reviewed lately?

Triggers of cost problems may be due to the following:

- Is there an increase in overtime?
- Is there an increase in general stores expenditures, such as gloves or tooling?

Triggers of personnel problems may be signaled by the following:

- An increase in attendance problems.
- Do personnel complain about rotating certain jobs?
- Are there certain jobs that people have difficulty learning?
- Has there been an increase in the number of suggestions concerning a particular process?

Triggers of other problems may be found in the following items:

- How does the maintenance log look? Is there an increase in maintenance problems and, more importantly, are they repeating?
- Are check sheets not being used?

Grasping the Situation

Another way to develop a problem-solving consciousness is to learn how to grasp the situation. First, the circumstances should be clarified by asking a few questions. What is happening specifically? What kind of documentation is there that something is wrong? The problem-solver must ensure that a problematic situation is more than a feeling. Can the feeling be quantified? Once a problem has been identified, he or she must ask "what should be happening?" and "what is the standard?" Can what is normal and expected be quantified? Get a firm grasp on the current situation and the standard.

The problem must be investigated to determine if there is one or several causes, then divided into three categories—man, method, and machine. At first, do not rule anything out. Potential causes should be discussed with all personnel involved to get their ideas. Go to the point of the problem to see it. One of the most important issues in solving a problem is completely understanding it, and it cannot be understood if it has not been seen. Many problems have been solved from a desk, only to resurface hours, days, or weeks later.

One method for locating the point of cause is to track the problem back to where it is no longer apparent. Then it is rechecked with a complete search of that entire location. Once the point of cause is located, managers are well on their way to solving the problem.

Seat Belt Case Example

This section presents an example using seat belts to illustrate how locating the point of cause can help solve a problem.

> B.J., a supervisor on a chassis assembly line, has an employee named Charles. Charles is normally a very good employee. B.J. has noticed lately that Charles is pulling the Andon more than normal (a 45% increase). B.J. talks to Charles' team leader and is told that "it's no big deal, he's just getting behind." As Charles is cleaning his area after the line goes down, B.J. asks how thing are going.
>
> "Terrible," he says, "I'm so mad I want to be put on another job."
>
> "Why?" B.J. asks.

"Boss, you know I try to do my best. I just can't keep up with the job anymore."

"Why?" B.J. asks.

"It's putting on that muffler with that stupid seatbelt always in the way. The seatbelt hangs down through a hole in the floorboard and it's in the way of where I need to attach the muffler."

"Show me," B.J. says. Charles takes him to the line and sure enough the seat belt is hanging down through a hole in the floorboard.

"Let's find out where this is coming from. Is that OK with you, Charles?"

"You bet, let's do it."

The two go back to the next car and the belt is hanging through the hole. They continue this process for almost 30 cars. At car 28 they see the problem. At car 29 they do not see it, nor at car 30. They have now located the point of cause at car 28. What is happening at car 28 that creates this problem?

B.J. asks Charles if he is willing to watch the operator on the process of car 28 and see if he can tell what is happening. The next day, Charles observes the operator. As the operator installs the glove box door, he is rubbing against the seat belt and causing it to slip down through the floorboard hole, which creates the problem.

After reporting back to B.J., a root cause is established (a new employee was not cautioned about watching for the seat belt because it was not included in his standardized work). The training took place and the standardized work was updated to include the seat belt caution. Charles' problem went away as suddenly as it had appeared.

THE 5-WHY CAUSE INVESTIGATION

Once the point of cause has been located, find the root cause of the problem. To find the root cause, use a tool called the "5-Why" cause investigation. Start by asking why is the problem occurring? Can the direct cause of the problem be seen? If not, what are the potential causes? How can the most likely causes be checked? How can the direct cause be confirmed? Then ask "why" and continue the cause investigation. Will addressing the direct cause prevent recurrence? If not, can the next level of cause be checked? If it is impossible to see, what is the suspected next level of cause? Then determine how to check and confirm it. Will addressing that level prevent future recurrences? If not, continue asking why until the root cause is found

The answers to the following questions will determine whether the root cause has been found:

- Can addressing this cause prevent recurrences?
- Is this cause linked to the problem by a chain of cause/effect relationships that are based on fact?
- Does the chain pass what is called the "therefore test?" Can problem-solvers begin with the final cause and after each reason say "therefore" to show a chain of circumstances that result in the original problem?

Once the problem-solver thinks he or she has discovered the root cause, ask "why" one more time to see if it leads to another problem not directly related to the issue being addressed. If this is the case, the root cause has most likely been discovered. For an example of a 5-Why investigation, refer to Chapter 4.

How do problem-solvers proceed when there is more than one potential cause? Once the potential causes have been divided into man, method, and machine, managers must challenge the potential causes. They must be challenged by using the facts observed, as well as the manager's own knowledge and experience. It is important to understand that analyzing for the root cause is not the same as identifying potential causes. When identifying potential causes, all the possible sources of the problem must be listed. When analyzing for the root cause, the most likely source(s) of the problem are tracked down to the root cause. An example of the full problem-solving process is described in the following windshield water leak illustration.

> This morning, water leaks are forming at the lower corner of the windshield on the passenger side of 23% of cars coming through the water test inspection. The

supervisor starts asking "why." If he or she knew the answer, a countermeasure could be instituted to fix the leaks. The supervisor asks, based on the documented facts of the situation and individual experience, "what are the potential causes?" There may be four distinctly different potential sources of the problem. Let's assume the following:

- The gasket around the windshield could be crimped during installation of the windshield.
- There could be pinholes in welds on the frame in the windshield area.
- Urethane sealer could be applied unevenly, leaving gaps in the sealer.
- The frame could be slightly out of alignment in the windshield area so the gasket is not making a firm seal.

These are all potential causes. One does not lead to another—each has to be considered separately. The manager should start with the facts of the situation to determine if they seem to contradict any of the potential causes. In addition, he or she must physically visit the site of the problem.

THE INVESTIGATION PHASE OF PROBLEM-SOLVING

In the previous example, the inspection report indicates the sealer and windshield gaskets on every car were checked in assembly. This information allows the supervisor to set aside those two potential causes for right now because they do not seem to be the likely problem. The facts indicate the sealer and gaskets are not the likely sources of the problem, but there remains a chance that they are. The manager returns to the list of potential causes and asks which of the two remaining causes is the most likely. The supervisor should base that decision on the accumulated facts and previous experience in the area.

The supervisor knows there are new inspectors being trained in the body weld area. Also, there was a similar pinhole problem a few weeks ago caused by the temperature of the welding robot being too high. Given these facts, the supervisor decides that pinholes in the welds on the frame area are the most likely cause. This cause should be tested first. Even though pinholes in the welds seem to be the most likely cause of the leaks, at this point the supervisor should not assume it is the root cause. This assumption has to be tested by getting more facts, which will provide a determination of the root cause(s) of the problem.

Other likely causes must be tested. This problem happened two weeks ago and this time the root cause(s) needs to be found to put this problem to bed. The supervisor should investigate to determine if there is a misalignment in the parts of the frame. If there is not, there is no need to pursue that cause further. The most likely potential cause should be run through the 5-Why sequence.

> There is a water leak at the lower right-hand corner of the windshield on 23% of cars this morning.
>
> *Why? Experienced body weld inspectors check cars leaving body weld.*
> There are pinholes in welds on cars leaving body weld.
> *Why? Based on their location, welds are traced to the robot that made them.*
> Robot #3 is burning through the center of welds.
> *Why? What is the temperature of robot #3 compared with the other robots that do not have a problem?*
> Welding temperature of robot #3 is too hot.
> *Why? Test thermostat and power circuitry of robot #3.*
> Thermostat on robot #3 malfunctions.
> *Why? Observe voltage meter.*
> Power surges affect welding robot #3.
> *Why? Check the surge protector on robot #3.*
> No surge protector on robot #3.

When analyzing for root cause, the supervisor continually narrows the focus of the search. As the search narrows, additional information on the remaining causes helps problem-solvers sort through the many possible explanations for the problem. As new facts emerge, it is possible to eliminate some explanations and concentrate on those that lead to the real or underlying cause. Root cause analysis is like working through the layers of an onion. This is completely different from considering potential causes, which branch out into many different directions.

Instituting Countermeasures

The PDCA model of problem-solving so far has dealt with elements in investigation phases. The problem has been identified, the cause has been analyzed, and a root cause has been discovered. Problem-solvers now must look for a countermeasure. The supervisor should make sure there is a distinction between a temporary measure and a countermeasure. A *temporary measure* is defined as an action taken to control a problem. The temporary measure may or may not address the root cause. The temporary measure is usually the first thing done to stop the problem immediately. It is the proverbial Band-Aid™. A *countermeasure* is a direct action taken to address the root cause of a problem and totally eliminate its recurrence. The following are some examples of countermeasures and temporary measures:

- Taping down the rear window molding until the window butyl dries is a temporary measure. The defect may go away, but the root cause has not been dealt with as to why the window molding needs to be taped down. Why is the molding rising?
- Grinding off a sharp tool edge to prevent scratches on an instrument panel is a countermeasure because there is a permanent fix for a specific problem—a scratch on the instrument panel.
- Writing standard work for the correct way to repair bumpers scrapped due to a process defect is a temporary measure because the reason, or root cause for bumper repair, has not been eliminated.
- An error-proofing device to ensure the correct drills are put in a transfer machine so the proper holes are drilled is a countermeasure because it stops a specific problem (incorrect holes drilled due to the wrong tools being installed).
- Adding a quality check in a downstream process to ensure the connection of a wire harness is a temporary measure because it is in a downstream process and it doesn't solve the problem. It only addresses the symptoms of a wire harness

not being connected. Can an error-proofing device be installed to ensure a perfect connection 100% of the time? If it were possible, it would become a countermeasure for this problem.

Generating Ideas

How are countermeasure ideas generated? The supervisor begins by conducting brainstorming sessions. As this is done, participants should think of as many countermeasures as possible that will either eliminate or at least prevent the root cause from occurring. The following guidelines are recommended for brainstorming sessions:

- emphasize quality over quantity;
- suspend judgement until later, and
- let one idea lead to the next.

The point is to avoid evaluating the issue prematurely. The brainstorming session should not be hindered by trying to decide what will work and what will not. Next, the list should be expanded by getting input from others. Also, supervisors have to consider the problem from the perspective of others, including:

- what has worked before or has worked elsewhere;
- what a supervisor or others affected by the situation would do;
- what someone with a different background or different skills would do (what would a new set of eyes see and do), and
- what countermeasures others would suggest.

Evaluating the Ideas

Once a list of possible countermeasures has been generated, it is time to begin evaluating them. A thorough evaluation is necessary before selection and implementation can occur. The evaluation of countermeasures includes two levels:

1. Screen the initial list of possible countermeasures and narrow it to a manageable number.
2. Conduct a thorough evaluation of the countermeasures on the narrowed list (to project and compare possible outcomes). Is it possible to use a simulation to investigate the effect of a proposed countermeasure?

The criteria for both levels of evaluation should include impact, effectiveness, and feasibility. In the initial screening, these criteria are used to evaluate a high general level. In a thorough evaluation, the criteria project and evaluate the possible outcomes of implementing the countermeasure from the narrowed list.

Common sense, experience, and an understanding of situational constraints impact the screening process used to narrow an initial list of possible countermeasures. The object of this screening is to identify and set aside any countermeasures whose impact, effectiveness, or feasibility is obviously questionable.

Initial Screening

For the initial screening, problem-solvers must ask themselves whether a particular countermeasure is likely to be workable in terms of its impact, feasibility, or effectiveness. Questions such as the following are frequently used to check the "face value" of potential countermeasures:

- impact—will this countermeasure create more problems than it will help solve?
- feasibility—is this countermeasure possible in this situation?
- effectiveness—can this countermeasure help achieve targets and goals?

Countermeasures that appear to be unhelpful, impossible, or undesirable are dropped from immediate consideration. The goal is to put a problem to bed. Just because a countermeasure is difficult does not mean it is impossible. Through this process, the manager arrives at a narrowed list of countermeasures worthy of a thorough evaluation. Note, however, that unworkable countermeasures are only set aside, not eliminated. They may be needed later if the narrowed list does not survive evaluation.

Developing and Standardizing Countermeasures

The next task is to develop a basis for comparing the countermeasures and selecting the best one to address the problem's root causes. The supervisor should try, as much as possible, to standardize countermeasures. This enables an easier countermeasure evaluation and repetition if needed. The following is a list of items that will help standardize countermeasures:

- Take steps to ensure that the countermeasure is permanently implemented at the process where the problem has occurred (point of cause).
- Track the countermeasure's effectiveness during the short-term, and ensure its continued effectiveness in the long-term.
- Include the new countermeasure as a "normal" situation.
- Check other processes in the group to determine if a similar problem is possible. If so, implement the countermeasure in all such processes.
- Communicate the results of the countermeasure to affected personnel.

CONDUCTING A THOROUGH EVALUATION

For a thorough evaluation, the basic question is: "What will most likely happen if this countermeasure is implemented?" The outcome of each countermeasure can be projected and rated on the basis of the three evaluation criteria. A grid such as the one displayed in Figure 16-1 can be used to develop ratings and organize them for comparison.

Effectiveness

Initially, countermeasures should be evaluated on their ability to achieve a target or a goal. To conduct the effectiveness evaluation (column 2 in Figure 16-1), the supervisor should ask the following:

- How well will this countermeasure work?
- Will this countermeasure be enough to achieve the target and/or goal?
- Will this countermeasure prevent a recurrence of the problem?

Temporary measures, which merely buy time, are not enough by themselves. They are often necessary to keep the line operating. Temporary measures may not, however, accomplish the ultimate aim of problem-solving, which is to prevent a recurrence of the problem (putting the problem to bed). The problem-solver must

Evaluation of Countermeasures				
Possible Countermeasures	Effectiveness (High/Medium/Low)	Feasibility (High/Medium/Low)	Impact (+/−)	Impact (High/Medium/Low)
1.				
2.				
3.				
4.				
5.				
6.				

Figure 16-1. Three evaluation criteria for proposed countermeasures.

continue to seek permanent countermeasures that address the root cause of the problem.

To complete the grid in Figure 16-1, grades of high, medium, and low are used to evaluate the effectiveness of a countermeasure as to whether it will achieve goals and/or targets.

Feasibility

Effective countermeasures must be evaluated to determine how realistic or practical they are to implement. When conducting the feasibility evaluation (column 3 in Figure 16-1), the supervisor should ask whether it is feasible to implement the countermeasure in light of:

- quality,
- cost,
- safety,
- resources,
- time, and
- management approval.

When evaluating feasibility, managers may use high, medium, and low grades to determine whether the countermeasure will be implemented.

Impact

Because production is so highly integrated, it is realistic to think that any countermeasure implemented will inevitably have some impact on other processes or areas of production. For this reason, supervisors would do well to use a two-step approach when evaluating impact (column 4 in Figure 16-1).

First, the supervisor should mentally project the type of impact a possible countermeasure might have on other areas/processes and consider whether the impact will be positive or negative. Based on those perceptions, a (+) or (–) is entered at the appropriate location on the grid. Then, the probable magnitude of the impact is considered. The supervisor should consider what effect this countermeasure will have on:

- the job,
- the team,
- other operations, and
- the company as a whole.

The extent to which others will be affected is specified by using high, medium, and low ratings (column 5 in Figure 16-1).

Once the evaluation in Figure 16-1 has been completed for each potential countermeasure, one usually rises above the others. Selecting the best countermeasure is no easy matter. Seldom is there a single, perfect countermeasure. Selecting the best countermeasure is a decision-making process that requires judgment and the ability to consider many factors at once. The following are some guidelines for selecting countermeasures:

- Keep both short- and long-term perspectives in mind. Think of what is necessary to keep the process running and what is needed to prevent the problem from recurring.
- Look for countermeasures that offer the best combination of results.
- The best combination of results often can be a balance of desirable and undesirable outcomes.

- It may be necessary to repeat the PDCA process to determine potential problems or anticipate uneven results from the implemented countermeasure.
- Before selecting a particular countermeasure, always ask: Does it really deal with the root cause of the problem? Does it really help achieve the target or goal? Will it prevent recurrence of the problem? Will it put that problem to bed?

This process provides a systematic way to evaluate and compare countermeasures. It is not, however, a scientific or foolproof basis for comparison. Always support decisions by facts and experience, and by physically seeing and understanding the problem. Temporary measures address the immediate steps in the cause analysis and help achieve the goal. Long-term countermeasures address the root cause and help achieve the goal of "putting the problem to bed."

THE "DO" PHASE

Up to this point, the problem-solver has been engaged in the investigation phase of PDCA process. Here is where the "do" phase comes in. In the problem-solving process, effective implementation cannot be assumed. Implementation must be planned, and the plan must be followed to determine if a countermeasure is effective. A plan is an important tool for both follow-through and communication in the system. In the do phase, the problem-solver develops, communicates, and executes an implementation plan. An example of a do procedure outline for the windshield water leak example follows:

- The root cause is no surge protector on robot #3.
- Develop a plan to implement the countermeasure to the root cause.
- Define the necessary actions.
- List and sequence the required actions.
- Specify who will do what, when, and where.
- Test the plan.

Ask "what if" to anticipate problems. Critical areas to question include:

- tight deadlines;
- costs;
- unclear responsibilities;
- possible adverse reactions;
- commitment by others to follow through;
- modifications to the plan if necessary;
- communication of the plan;
- use the problem report format (see Figure 16-2) or developing one unique to the situation;
- discussion of the plan with everyone involved with the problem, and
- execution of plan.

THE "CHECK" PHASE

In the PDCA system, problem-solvers do not simply implement a countermeasure and walk away. The "check" phase of problem-solving provides the means for checking both the implementation plan and the results of the countermeasures. First, the supervisor checks to see if the plan is proceeding as planned and on schedule. If it is not, he or she makes the necessary modifications. Once the plan is implemented effectively, the results of the countermeasures are checked. The following action items should be used in the check phase:

- Monitor progress of the implementation plan.
- Use the problem report to compare actual implementation dates to planned dates.
- Chart progress in terms of implementation milestones identified in the plan.
- Modify the implementation plan if necessary.
- Adapt or modify the plan to address changing conditions or situations.
- Report results and changes in the plan to appropriate groups, sections, departments, and personnel.
- Use a problem report to compare the actual impacts of countermeasures to projected impacts.
- Project and report results using measures of success that directly relate to the problem-solving goal.

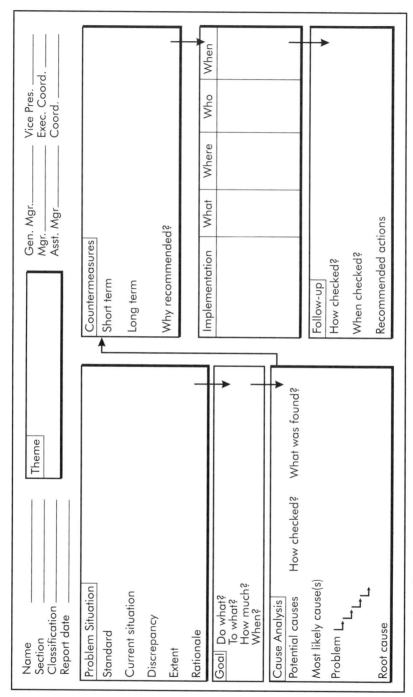

Figure 16-2. Sample summary report of PDCA problem-solving.

THE "ADJUST" PHASE

In the "adjust" phase of problem-solving, the task is to decide, based upon the results of the countermeasure's implementation, what should be done next. The results are evaluated first, then the appropriate actions are decided upon.

1. Evaluate the results.
   - Was the target or goal achieved?
   - Did the implementation plan work?
   - Will the countermeasure prevent recurrence?
   - Was the countermeasure partially effective?
2. Determine follow-up actions based on the results.
   - If the goal was achieved and it is evident that the countermeasure prevents recurrence, standardize the countermeasure and report the results.
   - If the goal was achieved, but preventing the problem is not ensured, repeat PDCA to find preventive measures.
   - If the countermeasure did not deal with the root cause of the problem effectively, repeat PDCA to identify other root causes and new countermeasures.
   - If the implementation plan did not work as planned, determine the root causes and identify countermeasures to use in future planning.

CONCLUSION

In lean manufacturing, the implementer is never satisfied with the status quo. He or she is constantly looking for continuous improvement. How does continuous improvement happen? Through constant problem-solving and putting more and more problems to bed, the implementer approaches his or her goal of having a problem-free process—a process that ensures safety, quality, and efficiency.

Pull Systems   17

Jim DeBold and John O'Meara

As companies begin changing their manufacturing strategies, one common notion is to implement a *"Just-in-Time"* (JIT) process of material procurement and distribution. JIT is one of the pillars of the lean manufacturing system. It is one of the most popular lean tools since it is one of the most tangible. The JIT concept focuses on producing the appropriate units in the required quantities at the necessary time and delivering them to the correct location. The results include a decrease in raw materials and work-in-process (WIP) inventories, increased cash flow, and the creation of available floor space for new business.

JUST-IN-TIME DEFINED

The JIT concept is simple: order materials only when they are needed. Schedulers plan production orders and purchase only the needed materials to manufacture the product. Fabrication departments and assembly operations produce only what is needed, then change to the next order. WIP is ensured to be 100% quality and moved directly to the next process. All parts are consumed in a brief period and cash is not invested in materials that will not be sold promptly.

If the company achieves JIT, managers can expect to see operational benefits that include:

- unnecessary inventories are eliminated;
- stores and warehouses are not needed;
- material carrying costs are diminished, and
- the ratio of capital turnover increases.

The bottom line is improved performance and significant impact on the financial performance of the company.

The material management group typically designs and manages the JIT process. Material managers take responsibility for all material procurement and internal movement within the plant. However, relying solely on a central planning approach that distributes production schedules to all processes simultaneously is not likely to result in a JIT process for all operations.

Potential Pitfalls

While the concept is sound, problems occur when there is a sudden change in the customer's schedule. It is nearly impossible for one central organization to effectively notify all operating departments of changes simultaneously. The effect is that fast-producing departments complete the items originally scheduled, but they are no longer required. Subsequently, raw material inventories may not contain the necessary quantities of parts now required, which causes the expedition of needed parts and costs a great deal in premium freight charges.

Likewise, if equipment does not perform as required, manufacturing operations may overproduce so they are not the cause of a system-wide failure due to the lack of WIP. Operations that require long changeovers tend to make the change worth it and run long batches of product to leverage out the downtime.

Ironically, while the intent of JIT is to reduce material buildup and minimize costs, the result is excess inventory within the plant, suppliers sending parts that are not needed, and costs incurred against products with nothing to bill against. To handle the situation, the company needs extra floor space for storage, material managers use expediters to satisfy schedule needs, and the company incurs overtime to make the parts now needed.

The Pull System as a Tool

A *pull system* is a key tool in lean manufacturing that enables effective just-in-time material management. The pull system makes JIT possible by transferring the responsibility of JIT from the materials management group to the production or operations group.

The pull system is a disciplined approach to JIT and provides an easy-to-use method that enables the company to maintain the needed discipline. The system provides an effective method of material movement management and production instruction, ensuring that the right parts are supplied, produced, conveyed, and presented only when required. The system accommodates schedule changes by quickly supplying the needed parts, while not overproducing parts no longer needed.

People often refer to the pull system as the Kanban system, which is incorrect. The *Kanban system* is a single element within the overall pull system. Kanban translates from Japanese as "display card." In many cases, this is exactly what the Kanban is—a card that displays the needed parts and instructs team members to move or make the item. Typically, the Kanban card informs the supplying process what the customer process needs.

TRADITIONAL MATERIAL MANAGEMENT

In traditional manufacturing operations, supplies are shipped in large quantities. This is usually due to the economics of transportation charges. To minimize costs, material managers order full truckloads for delivery, rather than smaller quantities that require more frequent deliveries. This inventory of raw material is then held in storage facilities and moved to operations departments in standard pallet quantities. The supplier and buyer usually decide standard quantities with little understanding of the actual need on the production floor. Stock that is not used relatively quickly is relegated to stores, where it must be handled, inventoried, stacked, and monitored until needed.

In-house production supply operations normally produce in large batch quantities. This minimizes the effect of changeover times and maximizes parts-per-day measurables. WIP not needed for that day's production creates cost in material handling personnel and equipment, such as forklifts and carriers (see Figure 17-1). In a traditional in-house supply operation, a schedule is sent to all production work centers from production control. In operations where large quantities are produced and stored, needed parts often

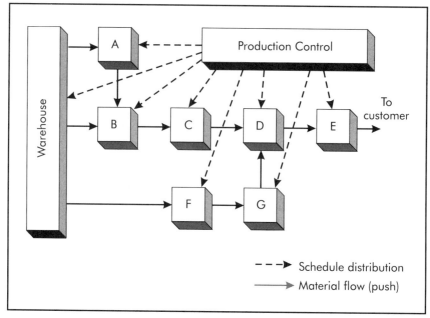

Figure 17-1. Traditional in-house supply operation.

cannot be located. This results in missed schedules, freight expediting, and unscheduled overtime.

PULL SYSTEM MANAGEMENT

A lean company uses a different method to communicate its production schedule and no longer tries to have all departments guess what the customer wants. Instead, the instruction to manufacture parts is controlled with the pull system. The schedule is sent only to the final production operation. This department takes (pulls) from the supplying operations the parts it needs to produce to the schedule. The supplying departments then replenish what was taken.

A signal is used to move or make all parts in a pull system. The focus is centered on supplying the next process with only what is needed, while ensuring high quality and on-time availability. Typically, each manufacturing department maintains a controlled "float," or buffer of parts that the customer may require. This is referred to as a "market." The volume and mix of the market are determined by the department's process reliability and its ability to respond to change. Purchased parts are controlled in markets that are set up for easy monitoring. Parts are ordered based on usage, the supplier's delivery confidence, and the economies of scale.

The last process in the operation, which completes the saleable item for shipment, receives the schedule and produces the parts required. To satisfy the schedule, material handlers or stock chasers retrieve the parts needed from the supplying processes to complete the order. This is done by means of a move signal.

The Move Signal

A *move signal* can be any means of sending information to the supplying process. In many organizations, the move signal is no more than a printed card covered in plastic. The move signal is sent to the supplying operation by the customer department, which indicates the name of the needed part, the number of parts needed, the storage location in the supplying department, and the delivery point in the customer department. Any information helpful to the material handlers is also included.

The Make Signal

With the parts removed from the supplying operations' markets and en route to the customer process, the schedule can be satisfied quickly. Once parts are removed from the market, a "make signal" is initiated to the supplying operation. This make signal informs the department that stock has been removed and must be replenished soon. The make signal indicates the part name, the quantity to be made, and the storage location for the part. Any information helpful to the producing department is included. Helpful information could consist of the items needed to make the parts, the machines, tools, or dies that make the part, or the item's quality documentation.

When an unforeseen schedule change occurs, it is handled nearly fluidly. There is no need to notify the entire manufacturing operation. Instead, the revision is sent to the final operation. This operation then changes from making what was scheduled to what is now needed. Material handlers take the move cards for the parts needed, go to the supplying operations, and withdraw the parts. Likewise, the make card is then initiated and the supplying process replaces what was taken. The make and move cards ensure that only the required items are supplied and/or produced.

As each operation becomes more reliable in terms of equipment performance and rapid tooling setups, the volume of parts in the markets (storage) can be reduced. The department should develop a systematic approach to reducing the buffer, thereby improving the company's cash flow. Team members should conduct an active problem-solving process to determine the new levels. The eventual goal or stretch objective is to completely eliminate the market and work strictly from make cards. When the process is predictable, the final process would send a make signal instead of a move signal directly to the supplying operation. The supplying operation then would produce parts in the quantity needed and deliver them directly to the customer, as shown in Figure 17-2. In a predictable supply operation, a schedule only goes to the final work center "E." The rest of the work centers get their instruction from pull loops.

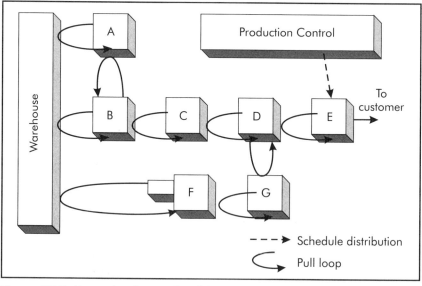

Figure 17-2. Example of a predictable supply operation.

GROCERY STORE EXAMPLE

One of the easiest ways to comprehend the pull system is to envision the way customers and items move through a grocery store (see Figure 17-3). In a typical grocery store, shelves are stocked with items that managers believe customers will purchase. This is based on historical data, seasonal influences, and the manager's experience. While managers can make general decisions as to what is most likely to sell and how much, they cannot determine exact amounts.

When purchasing the items to turn around and sell, store managers do not want to buy too much for fear of not being able to sell all of it. Likewise, they do not want to buy too little before the next delivery, or they will miss potential sales and possibly lose customers. In addition, they must be aware of and purchase the right amounts of product for seasonal events, such as candy for Halloween and turkeys for Thanksgiving.

Customers enter the store with an idea of what they need. Some have developed lists that indicate exactly what they want, while others choose as they go along. Either way, there is no scheduling by the store or the customers as to when the people come in, how many come in, or what they purchase.

Customers walk store aisles and withdraw what they want from the shelves. After each selection is made, the customer moves on to the next item. Routinely, stock handlers count how much inventory has been removed from the shelves. They obtain replacements from a back storage area and refill the shelves. As the amount in the back warehouse is reduced, the clerk makes note of when to order a new supply. An order is placed based on how fast the item is being consumed from the store's shelves, how much remains in storage, and how long it takes the supplier to deliver the product after an order is placed.

Some items cannot be stored for a long time, such as produce. For these goods, the shelves must be checked frequently and the amount in storage kept relatively small. These items are replenished frequently, or the result would be spoilage and loss of money. In this situation, the missing shelf items act as a move signal, which instructs stock handlers to move parts from the warehouse

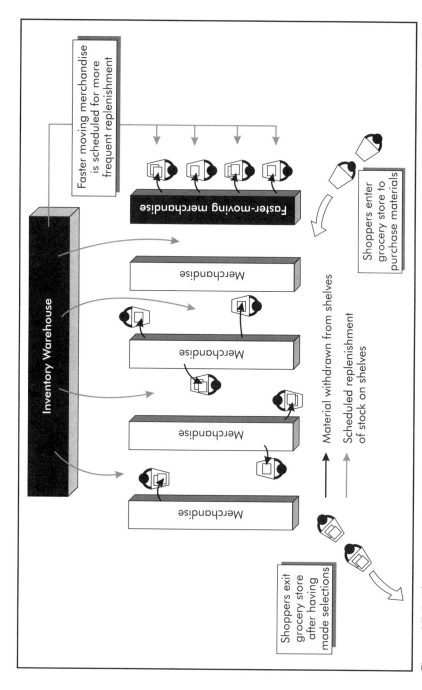

Figure 17-3. The grocery store pull system.

to the shelves. As items in back are depleted, a signal is sent to the producer to "make" more and deliver them to the store.

For items that can be kept longer than produce, such as cookies and cakes, checks of the stock level may be less frequent. Stock handlers may only check and replace these items once a day. Warehouse stock may be enough to last for the week and only be delivered on an as-needed basis. As with the higher turnover items, a move signal is used to request the transfer of items from the warehouse to shelves.

For small, inexpensive items that last a long time, such as soups and canned vegetables, it is not economical to buy small quantities from the producer, which would result in high transportation costs and frequent handling. Instead, the store stocks many of these items in the warehouse, places a significant amount on the shelves, and orders large quantities from the producer. The store may use a move signal from the shelf to monitor consumption. However, since the items are purchased regularly but are stocked infrequently, it is more economical for the store to use a "point-of-sale" system. In this system, the cashier keeps track of the consumption. This can be done manually or electronically. As the point of reorder is reached, the system issues a purchase order, or a make signal, to replenish the stock.

TOOLS OF THE PULL SYSTEM

Once an organization has committed to the implementation of a pull system, it must determine which tools will make it work. There are a number of tools available to the lean enterprise that make operating a pull system easy to understand and maintain discipline.

The most well-known tool is the Kanban. The Kanban contains instructions on what part is needed, its identifying codes (for example, part number, the quantity required, and storage location), and where the part is to be delivered. Usually, the card is protected from industrial elements, such as greases and handling, either by lamination or being deposited into clear, plastic pouches. The Kanban should be easy to make and to use.

Withdrawal Kanban

Withdrawal Kanbans are move signals that are circulated from the customer process to the supplying process. (An example of a withdrawal Kanban is shown in Figure 17-4.) Usually, there is a minimum amount of stock at the customer process, perhaps four hours' worth of parts. If the company assigns one Kanban for each standard container of parts, then the Kanban represents the quantity of one container. The Kanban remains in the container until the operator begins using the parts in that container. Once the operator begins to use the parts, he or she removes the Kanban and places it in a collection box. The operator then continues to assemble the items needed.

On an established schedule, a material handler or line feeder circulates through customer process areas and collects the deposited Kanbans. These Kanbans represent parts requisition orders to the supplying processes. Using the instructions on the Kanban, the material handler collects the parts indicated. For each Kanban, one container of parts is required.

Once all the parts are collected, the material handler then places the Kanban in the specified container and waits for the next delivery cycle. At that time, the handler delivers the parts to the station indicated by the Kanban and collects any Kanbans deposited since the last cycle (see Figure 17-5). The amount of stock at

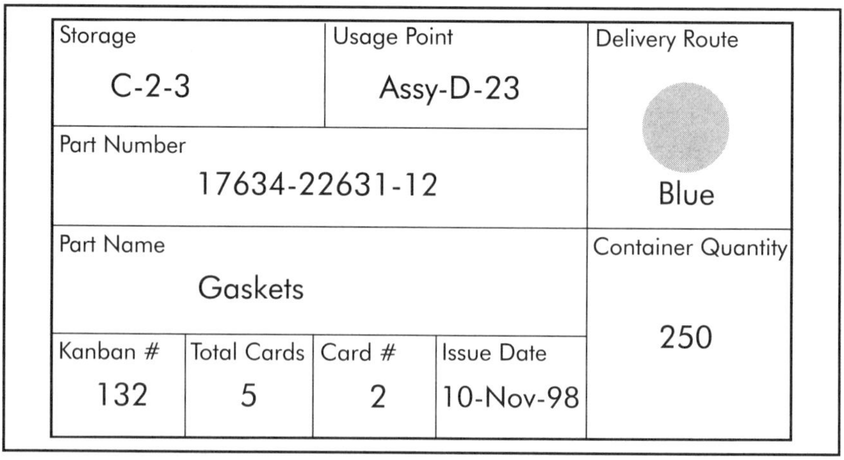

Storage	Usage Point			Delivery Route
C-2-3	Assy-D-23			
Part Number				
17634-22631-12				Blue
Part Name				Container Quantity
Gaskets				
Kanban #	Total Cards	Card #	Issue Date	250
132	5	2	10-Nov-98	

Figure 17-4. A typical withdrawal Kanban card.

Pull Systems

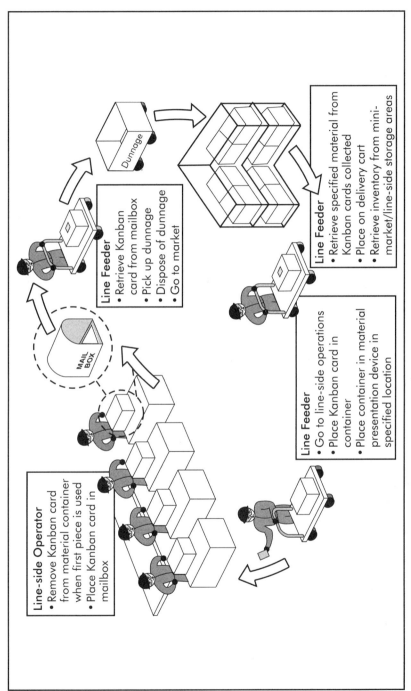

Figure 17-5. Sample delivery cycle using Kanbans.

the line is directly proportional to the frequency of delivery cycles. The more deliveries in a standard day, the fewer parts needed at the line. The fewer cycles, the more stock is needed line side.

Using withdrawal Kanbans has several benefits. It minimizes stock at the line, thereby reducing inventories, making floor space available, and allowing line-of-sight management. This reduction makes it possible to have a small amount of each part used at that station. As the customer requirement changes from one order to the next, only the parts being used are replenished. If the operator discontinues to use Part A and begins to use Part B instead, then there will be no further Part A Kanbans deposited. Rather, the operator will begin to deposit Part B Kanbans, and material handlers will retrieve only Part B containers.

The Supplier's Point of View

From the supplier's standpoint, it is important to set up a proper marketplace when using the withdrawal Kanban process. The market must contain any parts the customer may need. The amount of parts in the market is determined by the potential demand the customer places upon the process. For example, if a part is used routinely, there should be enough in stock to handle a worst-case scenario. For parts used infrequently, a small supply may be sufficient. Amounts should be determined by department personnel, based upon the known reliability of the manufacturing equipment and the ability to change to a different product within the department. One issue is paramount—the supplying process must ensure that only 100% quality parts are placed in the market. If not, the customer process will be forced to expedite parts when poor quality is found.

Production Kanban

A *production Kanban*, also referred to as a production instruction Kanban, is a make signal (shown in Figure 17-6) and can be used in several ways. The most common purpose for using production Kanbans is as a means to keep markets full. The production Kanban indicates when a predetermined amount of material has been removed from the market and initiates replenishment.

Supplying Process	Storage C-21-13			Requesting Process
Plastic	Part Number Plas-IB5X-143R			Assembly
Shipping Post No.	Part Name Console, Center			Receiving Post No.
Press #107	Quantity 25	Card # 1/3	Issue Date 11/13/98	Assembly Column 18E

Figure 17-6. Example of a production instruction Kanban.

In the market there is usually a method to monitor the amount of each part. Using visual management tools, each part can be identified with an established minimum and maximum level of stock. If the manufacturing operation is a stable process, then at no time should there be more stock than the maximum allows. Likewise, the minimum represents a safe stock level and the amount should never fall below it. To maintain the proper amount of stock (somewhere between the minimum and maximum level), the production Kanban acts as a trigger to tell the supplying process when parts are needed.

As in the previous example, in which one withdrawal Kanban represents one container of parts, one production Kanban represents the need to produce one container's worth of parts. In the market, each container has a production Kanban attached to it. The production Kanban has information that indicates what the part is, where it is made, where it is stored, and any other information that may be of importance.

When the material handler arrives at the market with the withdrawal Kanban from the customer operation, he or she removes the production Kanban from the container to be taken and places it in a drop box. The withdrawal Kanban is then attached to the container and the parts are delivered to the customer operation.

Sequencing

An employee in the supplying process is assigned the responsibility of regularly checking the production Kanban drop box. As the Kanbans are gathered, this parts-supply associate scans the cards and assigns them to the operation that produces the needed parts. In this way, the production Kanban acts as an instruction card, educating the operator on what is to be made, how much to make, and where to deliver the parts in the market. Once each order is completed, the operator attaches the Kanban and places the container in its assigned location in the market, thereby completing the cycle.

The production Kanban can be used in several different ways as a signal to make components. The easiest method is to produce the parts in the order the Kanbans are received (in other words, the operator receives the Kanban and begins to produce the specified item in the quantity indicated). Once that production run is completed, he or she begins to make what is indicated on the second Kanban. This order follows the "first-out, first-replaced" process.

When using the first-out, first-replaced method, it is important that the operator work on the Kanbans in the order they are generated. If the Kanbans are received in the order of A-C-A-B-A-B-C-A, the operator must produce the parts in that order. It is not uncommon for the operator to want to alter the sequence to minimize changeovers. The operator may attempt to produce the parts in the order of A-A-A-A-B-B-C-C. This enables all of the A parts to be completed before moving to the next part number. To the operator, and perhaps the supervisor, this appears to be the most efficient way to leverage downtime due to changeovers. However, in such a case, the system runs the risk of having an insufficient amount of C parts in the market during subsequent withdrawal cycles.

To be effective using the "first-out, first-replaced" method, the supplying department must be capable of rapid changeovers and using Kanbans in the order they were received. Operators and supervisors must understand the need to adhere to the process and maintain the ability to changeover rapidly.

Urgency Tables

Another method for using production Kanbans requires producing to an urgency table. An *urgency table* is a visual tool constructed to establish a priority for needed parts. The operator does not fill the Kanban in the first-out, first-replaced order. Rather, the Kanbans are satisfied by the priority shown on the urgency table.

The fundamental concept of the urgency table is that the number of parts in the market corresponds to the priority level on the table. The greater the volume of a single part in the market, the lower the urgency to replenish that part. Each part in the market is assigned a column on the table. The column is divided into four rows: the header on top, followed by green, yellow, and red rows. In each of these sections there is a peg onto which the Kanbans are stacked (an example of an urgency table is shown in Figure 17-7).

The header row consists of the part's name and any other identifying codes, such as a part number. It also indicates how many Kanbans are permitted in each of the green, yellow, and red sections of the column. Kanbans are placed on the board as they are collected from the market drop box in a predetermined order. The first cards go on the green section's peg until the maximum number indicated by the header is reached. Once the green row has reached the maximum cards allowed, the next cards for that part are placed on the yellow section's peg. Once the yellow section is filled to the maximum level, all remaining cards are placed on the red section's peg.

The number of Kanbans permitted in each section is determined by the changeover time needed to begin that particular part. The number of cards allowed in the red section is based on the absolute minimum the department wants on hand. This may or may

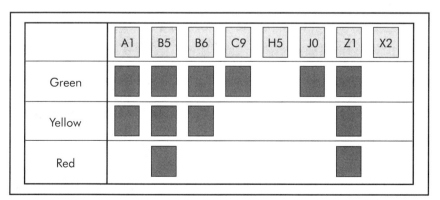

Figure 17-7. Sample urgency table for Kanbans.

not include any safety stock that the department has put aside. The green section contains the greatest number of Kanbans allowed to be taken before a changeover is considered. For example, if one container represents one hour's production at the customer process, but the supplying process can fill the container in 15 minutes, it is not worth the setup time to produce one Kanban worth of parts. The supplying department might decide that any production run of less than one hour is a poor use of the press's capacity. Therefore, the operator does not want less than four Kanbans before starting the changeover sequence. Accordingly, the header would indicate "four" in the number of Kanbans allowed in the green section.

The red section represents a mandated changeover. That is, the market inventory has dropped to a level where immediate output is required. The supplying department determines the minimum inventory allowed in the market, which allows the operator to respond while not shutting the customer's process down.

A card is placed in the red section when the green and yellow sections become full. When the first card is placed in the red section, the operator must respond with an immediate changeover. Subsequent cards show that the market inventory is becoming dangerously low and a line shutdown is probable.

The yellow section is based on the total Kanbans for the part in the market, minus the ones assigned to the green and red section. In other words:

yellow Kanban section =

total Kanbans in market − (green section + red section)

If a department determines that the market should hold a total of 10 containers, it must separate the 10 Kanbans into the green, yellow, and red sections. If the production team decides it will not run the press with less than four Kanbans, then green represents four cards. If the absolute minimum allowed in the market is two containers, then the red section represents two Kanbans. Therefore, the yellow section represents four Kanbans.

When the number of Kanbans on the urgency table is greater than four and less than eight, the operator has control over starting the process. Once the ninth Kanban is deposited, production must begin. When production begins, the operator must produce parts for all the Kanbans for that part on the urgency table.

Reducing Inventories with Kanban

As the supplying process becomes more reliable, department personnel should re-evaluate the volume of each component in the market. A common approach is to "tag" a certain amount of each part in the market. The tag indicates that these parts should not be used unless absolutely necessary. Material handlers are only allowed to use the untagged material.

The idea is that if the tagged material is not used, then it is unessential inventory in the market. Once it is shown that this material is not needed, it should be eliminated and the maximum level adjusted. However, if a situation arises in which the tagged material is needed (all untagged material has been consumed), then a manager must authorize the tagged material's use. Once the supply process has been stabilized, department personnel should then begin problem-solving activities to determine why the reserved material had to be used.

As the supplying process becomes more proficient at having the correct parts available, it may be possible to completely eliminate the market. In this situation, instead of the customer process issuing a withdrawal Kanban to the supply operation, it circulates a production Kanban instead. When the supply operation receives the production Kanban, it produces that part and supplies it promptly to the customer. Figure 17-8 illustrates the market between the supplying process and the customer process. The bottom half of the graphic shows a direct supply process to the customer. A minimum amount of emergency stock should be kept on hand in case of catastrophic failure.

Order Point Kanbans

Production Kanbans work well in operations where the model mix can be accommodated. Processes such as subassemblies, minor fabrication operations, and small welding procedures can quickly adapt to changes in demand. However, some operations,

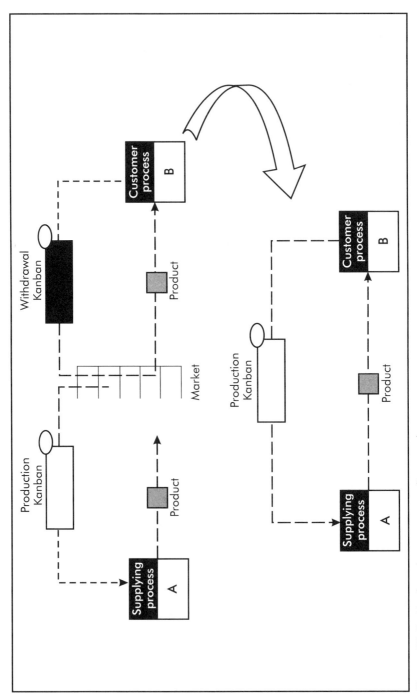

Figure 17-8. Markets between suppliers and customers.

such as metal stamping and plastics injection molding, cannot perform quick, sudden production changes.

When technology limitations make running smaller quantities unfeasible, a standard production Kanban is not appropriate. These types of operations require a special Kanban known as an *order point Kanban* (see Figure 17-9). The order point Kanban is initiated when a predetermined amount of material has been withdrawn from the market. When the material has been fully replenished, the order point Kanban initiates a tooling changeover and a production startup.

For example, in an injection molding operation, one molding press may be used to produce several different components. The die in the machine determines which part the machine produces. Depending on the changeover techniques used, the time it takes to change from one die to another may be significant. In such situations, a department's operations manager may not want to make changeovers without being able to produce a large amount of product. A changeover-related downtime then is leveraged over a greater number of parts.

If the changeover from one part to the next takes 60 minutes, the department may determine that the market must have six hours' worth of each part in stock. Note that six hours' worth of stock means the amount of parts the customer process can use in six hours, not the number of parts the supplying process can produce in six hours.

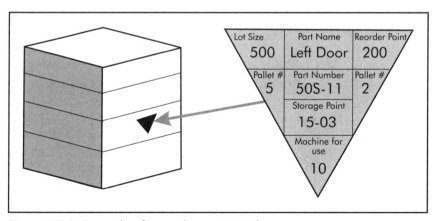

Figure 17-9. Example of an order point Kanban.

The molding department may determine that the maximum stock level is six hours and the minimum is 90 minutes. Subsequently, the replenishment point may be set at the three-hour stock level. With the market full, material handlers from the customer process take parts as needed. As the level of material in the market drops from six hours to five and down to four, no action is taken by the molding operation. As the amount of remaining stock reaches three hours, the reorder point is reached and the order point Kanban is circulated.

The order point Kanban is placed at the press and informs the operator which parts are needed, what machine and die to use, and how many parts are required. Since the machine is most likely making parts ordered by a previous Kanban, production of the new parts may be delayed. This is important to realize when planning the reorder point. While the machine and die are being set up, the customer process continues to consume parts from the market.

Once the machine is set up and running, the operator runs enough parts to satisfy the order point Kanban. The parts are then placed in the market at the position indicated by the Kanban, and the order point Kanban is placed where it can be used for the next cycle.

Manage Suppliers

Order point Kanbans can be used to order parts from outside suppliers. A market should be created that contains all supplier parts used in the manufacturing operation. Based on the volume of purchased parts and the location of their use, several markets may make more sense. Order point Kanbans also can indicate to the material manager when it is necessary to reorder parts.

Most suppliers ship parts in standard quantities determined by both the company's purchasing agent and the supplier's sales representative. These quantities typically cannot be renegotiated without significant cost penalties. It is common for the materials management group to reorder parts from the suppliers based on delivery frequency (parts are ordered based on the time of the month rather than actual need). This is done to ensure that shipping and receiving windows are satisfied.

An order point Kanban can be assigned to each part in the purchased parts market. While certain parts in the market, such as bolts, may be used in several places in the plant, they should be stored in only one location. A single location ensures that stock levels are controlled and parts are not over-ordered. For each part in the market, the order point Kanban should indicate what the part is, the supplier, the company's supplier code, the supplier's identification number for the part, the shipping quantity, and the reorder point. The re-order point is based upon the company's consumption and the supplier's delivery time once it is ordered. When determining the re-order point, the worst-case scenario must be considered. For example, assume the company has established a relationship with a bolt supplier. The agreement calls for the supplier to deliver the bolts two days after the order is placed. The supplier ships the bolts in pallet quantities of 60 boxes and does not ship partial pallets. Figure 17-10 shows a sample market replenishment process.

The company uses approximately 12 boxes of bolts each day, provided that there is no significant downtime in the operation. The company's management has instructed all operations to minimize the amount of purchased parts inventories.

In the past, the company ordered one pallet of bolts each Thursday afternoon and 60 boxes were delivered the following Monday. This was done regardless of the actual inventory levels in the market. The outcome of this practice was a severe fluctuation in the amount of stock, resulting in the need to count parts once a month and adjust the next order to correct the amount in the plant.

An order point Kanban can smooth out the ordering process. First, a reorder point must be determined. If the reorder point is placed at two days, or 24 boxes, there may be significant risk of late shipments. To accommodate this possibility, the company has decided on a reorder point of three days, or 36 boxes, resulting in a safety stock level of one day. If at any time the level drops below one day's worth of stock, the purchasing group must expedite the parts into the plant. This should be followed by problem-solving activities.

The order point Kanban method establishes a reorder point of 36 boxes and a minimum level of 12 boxes. To determine the maximum level for this part in the market, the company must examine

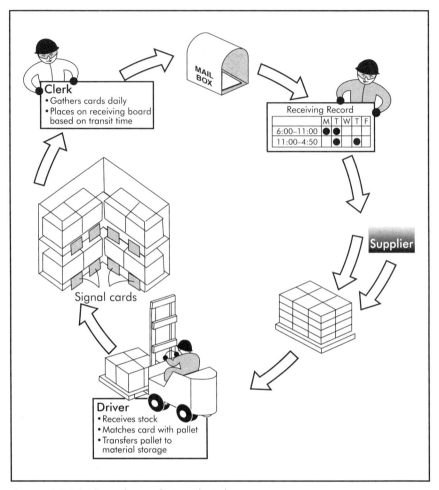

Figure 17-10. Sample market replenishment process.

a worst-case scenario. If for some reason the company's operation shuts down immediately after the order is placed, then the market has three days' worth of stock in the market when the order is placed. Since the standard delivery is 60 boxes, the maximum the market can have is 60 boxes + 36 boxes = 96 boxes.

The material replenishment cycle operates with material handlers bringing the withdrawal Kanban to the market and retrieving needed parts. As the quantity in the market decreases to 36 boxes, the order point Kanban is deposited in a drop box. The purchasing group collects the order point Kanban each day and places the order with the suppliers.

When the parts arrive at the plant, the delivery is reconciled with the order point Kanban. Both the pallet and the order point Kanban are delivered to the market. Any remaining stock from the prior shipment should be rotated with the new delivery to ensure "first-in, first-out" inventory management.

Broadcast Systems

Some situations make Kanbans ineffective when implementing a pull system. Parts stored in large, awkward containers are not easy to carry by hand. Parts with containers that are too heavy or fall outside of ergonomic limits may not be deliverable by a material handler with a pushcart. Such parts must be delivered with heavy-duty equipment, such as forklifts. They can be signaled by the Kanban system, but they require too many containers at the line. Also, delivery frequency may be too great to handle effectively. It may be better to use an electronic, or broadcast, system. This is sometimes referred to as a "call system."

A broadcast system is used when the logistics of a plant make Kanban less effective. Broadcast systems are commonly used for parts with low package density, that are awkward to handle, and/or fall outside of ergonomic limits. The *broadcast system* is an electronic program that allows the operator to send a semiautomated signal requesting part replacement. A forklift operator receives the signal and retrieves the needed material from its assigned storage area. It is then transported to the operator who made the request. Once the parts are delivered, the signal is extinguished and the driver returns to the signal receiving station.

This process works similar to a Kanban in that the operator initiates the request for more material. That is, the operator "pulls" the parts to the line. Once the signal is sent, a material handler responds by bringing only what is needed in a standard amount. If the schedule changes later, the operator will initiate a different signal that indicates the need for a different part. Figure 17-11 shows an example of the broadcast system.

It is the operator's responsibility to initiate the request, or call, with enough time to allow the driver to acknowledge the signal,

Lean Manufacturing: A Plant Floor Guide

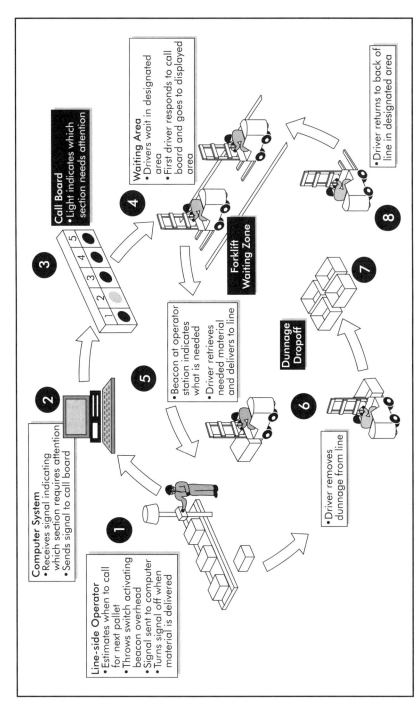

Figure 17-11. Sample broadcast system.

collect the material, and make the delivery. If the call is made too early, the material handler will have to wait at the delivery station until the first container is empty to place the new one. If the call is delayed, the material will arrive too late and a line is stopped. The appropriate time to make the call should be determined by a cross-functional team of production operators and material handlers. However, the probability of the last part being used as the material handler arrives is not high. A method for handling the few remaining parts when the new delivery arrives should be considered.

The concept of the broadcast system is simple. The design and details of such a system, however, are usually a function of need and available funds. A simple system may be no more than a light board illuminated by a switch at the operator's station. The board shows which area is requesting parts. Once the signal is viewed, it is the forklift driver's responsibility to identify what parts are needed based on the station making the request and the current model mix being run. That places the bulk of responsibility for proper parts delivery on the drivers.

A more elaborate system may consist of a signal button at the operator's station that sends an electronic signal to a computer monitor in a central waiting area. When the signal is transmitted, the monitor indicates which station is calling. If the funds are available during the system build, a computer system controlling the process can indicate needed parts by reconciling the station making the call, along with the model mix currently operating. It might also tell material handlers where the needed material is stored. This makes the system easy for everyone to understand.

The company should carefully address the range to which the broadcast system is developed early in the planning stage. Criteria for expanding the system, using more or less manpower, and future model mix demands must be considered. This makes any future expansions or changes easier to handle and accomplish. A back-up plan should be considered in case the system fails. There are many reasons why an electronic system might break down. Determining the cause and correcting it may take some time. The system developer should consider what will happen if the system fails and how material deliveries will be accommodated in the interim.

BASIC RULES FOR OPERATING A PULL SYSTEM

Regardless of what tools are used for the pull system, whether Kanban, broadcast, or some other device, there are certain basic rules that apply to their operation:

- Downstream processes should only obtain parts from upstream processes according to information provided either by the Kanban or another form of signal. If a Kanban is used, it has to be detached from the part's container in the downstream process and sent to the upstream process.
- The production process should only produce parts to the information provided by the "make" signal.
- If there is no pull signal (Kanban or otherwise) there will be no production and/or transfer of materials.
- If a Kanban is used, it should always be attached to the parts container unless it is in transit to order production or transport parts.
- The production process must ensure that 100% quality parts are produced before placing them in the market.

USING THE PULL SYSTEM FOR CONTINUOUS IMPROVEMENT

The pull system performs two major functions within the manufacturing environment:

1. It is a production control device, both in terms of material movement and information dissemination, and
2. It is an effective means of identifying areas for continuous improvement activities.

The pull system can be used to show constraints preventing stability and continuous flow. Once the pull system is running effectively, material movement throughout the plant should be smooth and predictable. Certain characteristics often manifest themselves when the system is working properly. They include:

- reduced radio traffic within the plant;
- forklifts operating at a slower, less hurried pace;

- a significant decrease in expediting parts in and out of the plant;
- an overall better attitude of material handlers, and
- production operators taking ownership of stock availability.

A properly operating pull system does not indicate that the process is completed. In any startup of a pull system, it is common to fill the markets with more material than is necessary. This is normal when making such a change to the material logistics program to safeguard against a customer shutdown. However, once the system is running and an adequate level of confidence has been achieved, it is time to systematically reduce the amount of parts in the markets by removing Kanbans.

Removing Kanbans

A plan should be developed to systematically remove Kanbans from the markets. Removing Kanbans is a tool of continuous improvement—it provides the opportunity to remove inventory (Kanban cards). Problems in the process will be discovered (inventory hides problems) when inventory is systematically removed. When Kanbans are removed, parts are not replenished to the market for those cards. Once the "Kanban-less" containers make their way to the customer operation, there will be no signal to replenish and the market will begin to thin out.

The first Kanbans to be removed should be those set too high. This is evident when parts always remain in the market. If some containers are never touched, then they are excess parts. The next wave of removal should focus on parts with the highest cost to the plant. Material managers should begin by removing one Kanban at a time and monitor the production operation. If the line continues to run uninhibited, the new Kanban level is established. This should be recorded and another Kanban removed.

If the Kanban's removal results in a line stoppage, then it becomes necessary to analyze what caused the shutdown. The materials department and the production operation establish a cross-functional team and perform a problem analysis of why that particular level is necessary. Items such as long changeovers, nonstandardized setup operations, raw material conveyance, unreliable machines, and/or infrequent line deliveries are all to be examined for their impact on reducing lead times for part production.

The problem-solving team uses the concepts of continuous improvement to isolate the root cause of the line stoppage, evaluate the uncertainty, and put corrective actions into place. Once corrective actions are in place they should be validated by once again removing the Kanban in question and monitoring the system for stoppage. Once the improvement activities are successful and the supplying operation is stable, the team begins to lower minimum levels of safety stock until it can be completely eliminated.

The long-term benefit of continuous improvement goes beyond the implementation of a lean program for moving materials. As problem areas are uncovered and lean corrective actions implemented, the entire manufacturing operation becomes more stable and predictable. Ultimately, with patience and perseverance, one-piece flow is possible, which results in reduced operational costs and much shorter lead times.

EXPECTED BENEFITS

When the company's managers decide to implement a pull system within their operation, they usually have a specific objective. More than likely, the goal is to reduce bottom-line costs by reducing inventory and/or correcting missed shipping windows. Regardless of the reason, common outcomes are seen with the proper implementation of the pull system.

Improved Material Flow

The first outcome is a smoother flow of materials from the receiving dock to the operation. The pull system provides a more proactive way to obtain materials. This is evident when expediting parts from storage areas to operations is nearly eliminated. Also, if the company's material handlers use radios to indicate trouble areas, there is a reduced amount of radio traffic.

Inventory Stability

Once the pull system is up and running, there may not be an immediate reduction in inventory. In fact, inventory levels may increase slightly because most companies do not significantly reduce their in-house inventory when a new program starts. Materials managers may fear that lower stock levels are too risky and jeopardize the plant's ability to satisfy the customer. However, the company should see stabilization in how much inventory is in the building. Fluctuations in the amount of stored parts should be minimized. As confidence in the system increases, a planned reduction in parts storage should soon be achieved. In traditional manufacturing systems it is not unheard of to see inventory reductions as great as 80%, resulting in more production room rather than storage space.

Reduced Freight Costs

If the plant has been experiencing a great deal of expediting, there should be an immediate impact on the cost of premium freight. In many situations where the company must have certain parts to run, it ends up paying all expediting costs. However, with the pull system there should be no need for expediting, provided that the signals used to reorder stock are honored. If the supplier is unable to provide parts in the agreed upon time frame, then the cost to expedite should be the supplier's responsibility, not the company's.

No Waiting

The greatest immediate benefit of a pull system is that no operator in any of the production processes has to wait on material deliveries. If there are areas in the plant where operations stop due to temporary material shortages, pull systems will eliminate these occurrences. Since all parts are at the production process in some quantity, they are always available. If Kanbans are continuously circulated, any parts being used by the operator will be replenished quickly. Therefore, the operator never has to wait on parts. Since operators no longer have to wait on materials or the correct parts, they are more productive—they are able to produce more parts per standard shift. As a result, there should be a reduction in unplanned overtime.

PULL SYSTEM CONSIDERATIONS

The pull system operates efficiently provided certain disciplines are followed. There are many stories in industry about the failures of pull systems. The universal response is "that program won't work here." However, investigation concludes that failure is due to the lack of planning and proper management of the system. The consequence of inadequate planning is that production and material handling personnel now have reason to resist any further attempts to make changes to the plant's operating program.

The key to the program's success is understanding the actual customer. An operation that sends parts to a market must understand that the market has become the customer, not the operation that withdraws parts from the market. The market is managed by a series of control points referred to as the minimum and maximum inventory levels. Maintaining the float between these levels is important to ensure the availability of parts.

Use the Time Allotted

In some operations with strict cost controls, a manufacturing manager may decide that replenishing the market is not critical. Instead, since the production operation is functioning, there is no need to keep the supplying process operating. Therefore, the manager does not approve the time needed to properly replenish the market. Unfortunately, the outcome of a market with a low inventory level is that the customer process starves for parts. Even if the supplying operation resumes production the next day, there has been an interruption to the flow of material. The result is the customer operation goes down and cannot produce saleable items, and the supplying operation must use overtime to make up the parts needed. The company experiences higher costs with no product to charge it against.

Material Handling Issues

When preparing to launch the pull system, company managers must understand that there will be a shift in the workload with which material handlers must adjust. Since all the parts are to be delivered more frequently, material handlers will be carrying less parts, more often, to the line. This increases the amount of distance the handlers must travel in a day. The company should be prepared to add personnel if necessary—it adds cost but it should be significantly less than the obtained savings. If the company finds that the current number of material handlers satisfies the new system, then there were probably too many in the old system. The number of material handlers should be examined for excess. If there are too many handlers, they can be moved to areas where their efforts will provide more reward for the organization.

Maintain Delivery Cycles

Another potential problem area is that material handlers do not understand that they are a service organization dedicated to the production operation. They must realize the importance of completing each delivery cycle as designed. Failure to do so results in fluctuations in the material process, and the customer process does not have sufficient parts at the operation to continue production.

In a pull system it is important that delivery cycles be maintained throughout the day. If the material handler misses one cycle, he or she may be able to correct it on the subsequent round. However, the burden of material containers is quite high and that hinders the completion of the cycle on time. Likewise, since the handler missed the initial cycle, a production Kanban is dropped from the market during that round. There will be twice as many production Kanbans deposited in the next round, which may overwhelm the supplying operation.

Proper Training

It is important to spend the time to properly train material handlers on their role in the process. They are the ones who make the program a success or failure. Some plants have stock personnel that spend the morning supplying the line with enough parts to last the day. Then they can spend the remainder of the day doing personal business or perhaps even going home. In such operations, stock handlers will not welcome a system that requires continually circulating the operation.

If the plant is unionized, it is likely that the material handlers are the seniors of the work force. Material handling is a job of high esteem because it does not require much in the way of physical work. The handler remains on the forklift most of the day and travels throughout the plant. This mobility makes the material handler an integral part of the plant's informal communication network. It is also the reason that union officials will not support a program the material handlers do not want. If he or she supports a program by management that works against the material handler's desires, the material handler can become very vocal in his travels and cost the union officials votes in the next election.

Parts Must be Perfect

The program will not be successful if supply operations do not ensure 100% quality parts in the markets. This includes parts from internal and external sources. If the customer operation receives parts that are unusable, considerable time and resources are spent expediting parts to the operation. If this continues to be a problem, then the customer operation has no choice but to inspect all parts before they are used. This adds the costs of personnel, handling, floor space, and double handling. The supplying operation must take responsibility for all parts it produces and work to guarantee quality.

Use Triggers

The easiest way to make the pull system fail is to not initiate the trigger mechanism. Whether using a manual system (such as a Kanban) or an electronic program, if the operator does not make the "pull" when required, parts will not be delivered on time. The result is a line stoppage. In companies where material control is the exclusive responsibility of the stock chasers, line operators have no responsibility for the correct parts being available. In a pull system, operators must take charge of requesting the material. This places a burden on them that they may not want. Training must be done to ensure that the pull is made properly and on time. It should be noted that, with little exception, most operators and material handlers have expressed that they preferred the pull system over the traditional programs once it was up and running.

COMPANY CULTURE

The attitude of managers has the greatest influence on how the work force will accept the new material replenishment system. The philosophy that "the line must *never* stop" is a cornerstone concept in many manufacturing operations. A common understanding is that if the company is not making parts it is not making money. This becomes a major source of apprehension for all personnel.

As long as the people are held accountable for nonstop production, they make every effort to safeguard that the line continues to run, whether it is with the correct part or not. They take precautions so that if the line does go down it is not their fault. Managers must understand that this philosophy results in excess stock and shoddy repair methods. It is important that the company be honest in recognizing that they may be broadcasting this message through actions, even if they speak differently (in other words, actions speak louder than words).

To prevent line stoppages, operators want large amounts of parts in their area. This ensures that regardless of the response time of the material handler, the operator has parts to keep the line running. That way, if the line does stop it is someone else's fault. Also, high stock levels act as a barrier to the main aisleways and prevent managers from scrutinizing an operator's actions.

Material handlers may like large amounts of inventory to provide a greater amount of "warning" to supply the line. Occasionally, the material handler provides the line with the stock needed and stages material close by. This allows the material handler to quickly feed the line when the operator's material runs low and safeguards the material handler from being the reason the line stops. Once he or she uses the staged material, a new batch is gathered and placed nearby.

Production control departments tend to order material in excessively large quantities to ensure availability and minimize transportation costs. However, the burden that a large amount of material places on the manufacturing floor is typically unseen from the production control office but must be endured by the floor personnel.

TRAINING

Lean manufacturing and Kanban pull systems are a big cultural change for all companies that currently do not use lean tools. For a change of this magnitude to take place, communication and teamwork must be emphasized. This section deals with some of the common change management issues, concerns, and best practices in industry. The time it takes to deploy depends on the audience. Some people can read about a process and gain all the confidence they need to execute an implementation. Others need to see a lean system functioning to appreciate and understand it.

Commitment from the organization's management is essential for the program to succeed. Once this is attained, a pull system implementation team must be formed to lead the effort. This team should consist of competent people who are motivated to succeed (not problem personnel whose manager wants to pass them on). Honesty is important here. The quality of the team is reflected in the quality of the program. People with complacent attitudes will not be committed to changing an existing system.

The team should consist of employees knowledgeable in production, materials management, labor relations, ergonomics, skilled trades, and related fields. Leadership commitment is demonstrated by allowing workers to leave their current positions and form the implementation team. Once the team is formed, it is important that team members receive as much training as possible. Training may consist of reading written materials, attending conferences in lean manufacturing, and/or benchmarking companies that are making the lean transition. It is important that the team becomes knowledgeable and develops a vision of what it is trying to achieve.

As the implementation team develops the new system, the remainder of the plant should be made aware of what is to come. Early awareness helps alleviate the apprehension of line personnel as changes move forward. At these preliminary awareness sessions, the implementation team should be introduced to the production workers, a means of communication should be described, and the vision explained. This helps to reduce nervousness so that when line personnel see the implementation team in their area they will be less likely to look for hidden agendas.

Training is the key to launching the program successfully. No matter how good the planning, if workers are not properly trained in concepts and operations, the program will not succeed. The training program should continue to evolve. Problems that surface during the initial training classes should be dealt with in future revisions.

Line Personnel

Once development issues are underway, the implementation team should develop training programs based on the knowledge they will have achieved through the pull system's development. This training should focus on targeted groups. For example, line personnel, such as production operators, should understand the system but there is no need to provide details of the system's development. Line personnel have little interest in the company increasing the number of inventory turns. The question they will want answered is "how does this affect me?" Training for line personnel should be simple, brief, and often repeated. They must know the basics of the system and the role they play. They should be assured of their importance in the process, the importance of the system, and that losing a Kanban will not cripple the system. Likewise, they should be aware of the consequences of intentional alteration of the process.

First-line Managers

The team must provide training to first-line managers on the system's dynamics and potential benefits. Line managers need to understand that their productivity is reliant on the successful delivery of stock. They should be aware of breakdown indications, such as Kanbans seen in places other than attached to stock or in drop boxes, unusually high or low levels of stock, or long delays in material handlers repeating their cycles. They should have a person in materials management to contact when problems occur.

Material Handlers

Training for the material handlers is more in-depth than for line employees. Material handlers are the ones who will make the system work. They must have a complete understanding of how the pull system operates and how it was developed for that specific facility. It may be helpful to use a simulation, or model, of the pull system to demonstrate the program's operation. Likewise, visits to other plants using pull systems are helpful, particularly if material handlers can talk with their counterparts. Since production operators are their direct customers, material handlers should get the same training that the line workers do to comprehend what the "customer" understands.

Material Managers

Managers from the materials control group must not be overlooked in the training process. Ideally, they will be involved in developing the process from its inception. While managers may not know the day-to-day details of each decision, they must understand the intent and logic used to develop stock levels and delivery routes. This makes them more capable of responding to issues as they arise. They must understand why the company is adopting a pull system. If the pull system is important, they will handle issues in a lean way rather than reverting to past, traditional practices.

ENSURING SUCCESS

While no one can guarantee that the pull system will work successfully in every operation, certain precautions can be taken to minimize the chance of failure. The main cause of failure in any transformation is under-communicating that a change is about to occur. It is imperative that the system not be developed "in a vacuum." Organization leaders must provide clear reasons to all those affected as to why the pull system is being adopted. Whether it is to improve cash flow, ensure market share, or save the plant from closing, the people who will use the pull system must know why it is being implemented. Employees must understand that the inconvenience of change is less than the danger of staying the present course.

Implementers must make sure they understand where the potential for failure is when planning pull system logistics. Losing cards is the most obvious way for the system to break down when using Kanbans. While one or two cards may not have an impact, a continuous loss breaks down the system.

The implementation team must accept liabilities and understand why losses may occur. Some employees in the production operation will intentionally discard or ignore the triggers just to demonstrate their contempt. To help minimize this possibility, the team must provide adequate training and explain the importance of their role. Likewise, managers must be trained to see the telltale signs of unusually low line-side inventories. If it is determined that someone is working to thwart the system, appropriate action must be taken.

Make Kanbans User-friendly

More times than not, lost Kanbans or failing to initiate the broadcast system is due to an oversight. It is not uncommon to find Kanbans piled up on a table next to the operator or in the dunnage. This is due mostly to the fact that mailing the Kanban is not a cyclical operation. It only happens when a new container is opened. This may be infrequent, depending on the container's quantity. The implementation team should place the deposit box, or mailbox, as close to the operator as possible. The underlying notion is that depositing the Kanban should be as minor of an inconvenience as possible. A rule of thumb is that the receptacle for the Kanban should be no more than a step away from the operator. Call buttons should be within arm's reach.

Do Not Overestimate Material

Another area of possible failure is overestimating how much material delivery personnel can handle in a single cycle. If there are a great number of Kanbans deposited during each cycle and there are few material handlers, the volume of requested parts may overburden the delivery. In such a case, parts are delivered to the line late and the process is continuously trying to catch up. Likewise, if the number of Kanbans deposited during each cycle are few and there are more than enough material handlers, then the manpower is underutilized. Inefficient manpower results in added waste by having people wait on the process. Worse, it is likely that the material handlers will skip cycles since the demand is relatively insignificant.

Create a Solid Delivery Process

The implementation team must set up the delivery process in such a way to allow all materials to be delivered on a timely basis. Many times the team only considers the actual delivery of stock to the line. It is important to recognize that the delivery cycle includes not only supplying materials to the line, but picking up deposited Kanbans, collecting dunnage, retrieving parts from the market, and preparing for the next cycle. Other things to consider are the distance the material handler must travel, the rate of part consumption at the line, the location of parts in the market, and any other duties assigned by the material manager.

Design Simple Processes

The implementation team typically is focused on completing the program, getting it running, and then moving on to the next project. One issue often overlooked is that material handlers who were there when the system kicked off will not always remain in those positions. People transfer in and out of the materials group, and others will be sick or on vacation. It is imperative that the system is set up to handle new workers and temporary replacements. Problems occur when new members come into the materials group. They do not understand how "pull" operates and the discipline that is required. The effect is that the tenured members must make up the deficiency, which leads to overburden.

The pull system should be designed as simply as possible. The process must be standardized and exceptions should be an infrequent occurrence. If the development team finds itself making a lot of conditional rules, then the program will fail due to complexity. In this situation, a development team can be a group that provides input on the outline of the system or one that executes or implements the plan.

Standardized Work Sheets

Standardized work sheets should be used at all material-handling wait stations. For personnel responding to Kanbans, the work sheets should indicate delivery cycle times (a key to understanding the information on the Kanban) and a resource to call for help. For the broadcast system, a standardized work sheet should detail where the call stations are, where the material is stored, and what to do with dunnage. The trigger process should be incorporated on all the standardized production work sheets.

Visual Aids

Visual tools indicating delivery routes will assist new members. Whether the route is marked on the floor or signs are hung from building columns, visual aids keep delivery personnel on track and ensure that they do not wander into the wrong areas of the plant. Along with route signs, stopping points can be identified. Instead of the driver looking into the production area to find Kanban cards, a simple line on the aisleway would indicate where to stop.

Identify Countermeasures

To safeguard the success of the pull system, the implementation team should envision possible failures and take appropriate action. Potential problem analysis (PPA) should be conducted on a continuous basis in regard to what could go wrong, the probability of the occurrence, its impact on the system, and possible countermeasures. For example, if the delivery system uses a battery-powered tugger to make deliveries, what happens when the last shift neglects to charge the unit and the power fails in the middle of a shift?

CONCLUSION

The pull system is probably the most recognized component of a lean enterprise. It can greatly improve material flow and free wasted floor space. Supplemented by the other lean initiatives, such as quick changeover, error-proofing, total productive maintenance, etc., pull systems can greatly reduce lead times and ensure customer satisfaction.

Pull systems, however, are not as simple to operate as they may appear. Like any operation, the pull system is not without its problems. Failure to properly calculate material levels, lost cards, missed signals, and late deliveries will cause the system to fail. The challenge is to adapt the program to the operation in such a way that employees will take ownership of the program. Proper planning and execution is critical to the program's success or failure.

Section IV:
Lean Applications

Greenfield Site Implementation   18

(new plant)

By Richard Dixon and Charles Robinson

The best opportunity for deploying the principles and tools of lean manufacturing is to begin with a new or empty facility. This is known as a *greenfield site* implementation. In a greenfield implementation there is nothing to transform and no habits to unlearn. Equipment moves are easily made on paper, as opposed to physical moves on the plant floor (although some engineering habits may need to be changed).

The early focus for those implementing lean at a greenfield site is on engineering (both product and process) instead of the plant floor. Most plant designs are based upon previous plant designs. A previous mass-production design will not enable a lean operation.

Nearly all projects go through the following phases on the road to operability:

- need identification/concept definition,
- conceptual design,
- program approval,
- detailed design,
- construction,
- startup,
- ramp-up, and
- normal operation.

If the project is to be lean, each of these phases must include specific action items to ensure that the plant is "born lean." This chapter describes the actions that need to be performed and when to perform them to ensure lean implementation and operation.

IDENTIFYING NEEDS, DEFINING CONCEPTS

Companies are constantly seeking new markets for existing products and for new products that will either satisfy the needs of the marketplace or create a new need. Assume that a company's product developers have created a new product or a significant improvement to an existing product. Or, perhaps marketing has uncovered a hidden demand for a product already being manufactured elsewhere. Now, senior managers have decided to conduct a feasibility study on the construction of a new facility in a new location, or a new line to be constructed at an existing facility.

Is it too early to begin thinking about lean? Definitely not. What can lean contribute to the feasibility study? The answer is a surprisingly large amount. A plant "born" lean can be less expensive from a capital project standpoint. Lean manufacturing requires less space for work-in-process inventory and has a different organizational structure, with more emphasis on the operator. Feasibility studies using lean guidelines will identify the most cost-effective methods for achieving the company's goals. Company managers will then determine whether to go forward, and in which direction the expansion should proceed.

CONCEPTUAL DESIGN

Lean activities to be performed during the conceptual design phase focus on making sure the staff understands lean principles, the company benchmarks to world-class standards, the layout minimizes monuments (inflexible, large-capacity machines), sufficient training is planned to launch a new system, and lean metrics are defined as the operating measurables.

Engineering and financial staffs must have a good understanding of the principles of lean to develop a valid feasibility study. Using mass-manufacturing assumptions will yield a performance inconsistent with the projections of the feasibility study. It is important that all principals be in agreement with the assumptions used in the study (with the awareness that local conditions may warrant adjustments to the assumptions).

Financial Issues

Financial assumptions should be based upon world-class standards of cost, quality, and time instead of organizational bests. This means the company must benchmark recognized world leaders and establish comparable goals so the company remains or becomes competitive. No long-term value is achieved by benchmarking less than the best.

Planned capital expenditures must minimize the number of monuments and catch-up operations in the proposed layout. Monuments result in excess inventory and all the associated negatives. In other words, do not plan on a three-shift operation of a high-capital-expenditure item to keep up with a two-shift operation of a less expensive operation (for example, using three shifts to build inventory that is used up in two shifts of assembly operations). Plan capital requirements based on equipment and operations meeting the Takt time of the customer's requirements on a shift-by-shift basis, and purchase right-sized equipment.

The ability to effectively run a lean manufacturing system in a new facility is directly proportional to the commitment of running waste-free operations and the success of training given to employees. Conceptual designs require a substantial amount of training expense per person, in addition to the expenses for training materials that are different from a traditional plant-launch curriculum. Cutting costs in training will guarantee a less than successful launch.

Metrics

The final core element in the conceptual design phase involves establishing a set of metrics from which to determine the "leanness" of the design. Typical lean design metrics may include:
- product travel distance for manufacturing;
- target model changeover time;
- percentage of capital budget spent on value-adding equipment—a stamping press is a value-added capital item, a conveyor is not;
- percentage of floor space devoted to value-added activities—process areas are value-added, a warehousing area is not;

- total planned value of maximum in-process inventory;
- target mean-time-between-equipment-failures, and
- flexibility indexes, such as the cost and time to add 25% capacity, per-unit cost differential at a run rate of 50% of capacity, percentage of equipment that can change over from one model to the next within cycle time, and percentage of machines that can be moved from one location in the plant to another within eight hours.

An implementer's conceptual design should be based upon core lean metrics and the associated implementation and operational costs to support them. Benchmarking will provide a good indication of what the core metrics should be. Most lean companies are willing to give guidance regarding the costs incurred to be "world class."

PROGRAM APPROVAL

Program approval is the project phase in which the commitment is made to proceed with the capital project. If the project is to proceed with full approval from decision-makers, as well as having the highest probability of being lean, certain standards must be maintained.

Many games are played during the program approval process to ensure that the program gets approved. Those who present the project to decision-makers have devoted much time and effort toward developing the project to this phase. Project approvers are wary of inflated promises of performance and bloated budget projections. The normal methodology in this situation is to cut capital budgets as a challenge to design teams, while accelerating time frames. Designers are often judged on total installed costs, leaving operations to live with a design that results in less than optimal total life cycle costs. This process normally results in unbalanced, inflexible, and unreliable processes that are difficult to operate and impossible to maintain at a level envisioned by the program approvers. To correct this problem, the basic culture of the approval process must be changed. The approval should be based upon different parameters other than simply a return-on-investment projection.

DETAILED DESIGN PHASE

During the detailed design phase, the focus must be on preventing waste. It is the only opportunity a facility will have to *prevent* waste. After the bricks and mortar are complete and the processes installed, the focus turns to waste reduction. A good mantra to follow is "refuse to include waste during the planning stage and fewer resources will be needed to reduce it during the operations stage."

Figure 18-1 is an example of a typical mass-production plant layout. The areas with thick gray borders indicate value-added operations, and thin-bordered areas indicate non-value-added operations. Some activities within the value-added areas may not be value-added. If process flow is followed (indicated by the lines), there is a significant amount of material handling in the plant. Does material handling add value to the product? No, so the plant should be laid out in such a way that material handling is minimized (see the lean layout in Figure 18-2).

Compare the differences between the two layouts. The lean layout:

- uses one-piece flow, a system where only one piece of in-process material separates operations;
- has virtually no material handling;
- has operations positioned in a U-shape;
- has only two storage areas (marketplaces);
- has much less non-value-added space;
- uses Kanban cards (squares with black circles) for material replenishment to the shear and the assembly line;
- results in a much smaller plant footprint, thus saving a significant amount of space and capital;
- uses scheduled logistics to eliminate the need for extra shipping and receiving doors—live load/unload;
- has no repair areas—material is repaired within the station or it is scrapped;
- has no inspection station—each operator is responsible for his/her own quality, and
- has right-sized machines to meet customer demand.

The comparative diagrams show that bricks and mortar can be reduced substantially, resulting in significant investment savings.

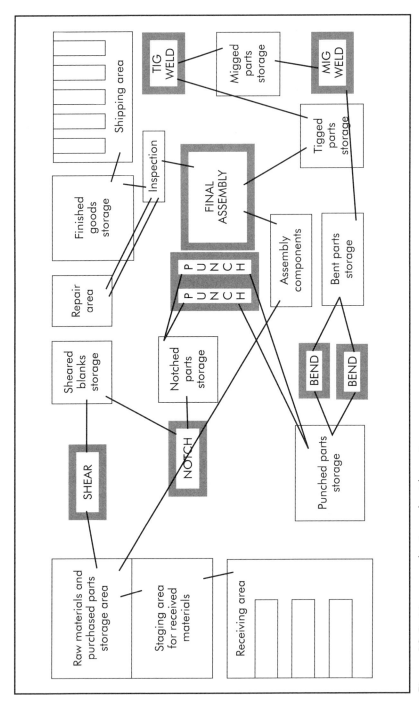

Figure 18-1. Mass-production plant layout.

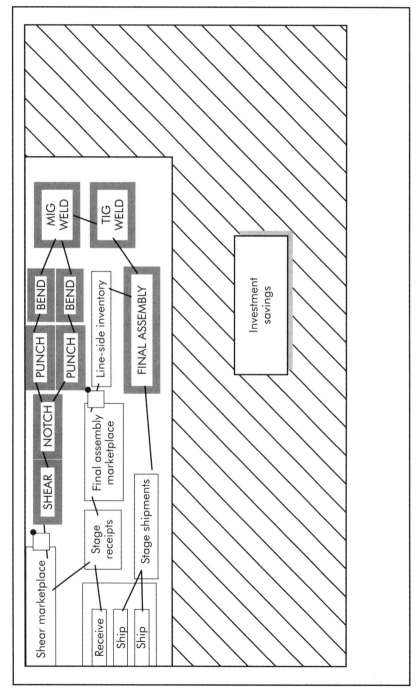

Figure 18-2. A lean layout.

Right-sized equipment reduces equipment investment in both quantity and capacity. Services can be reduced as well. For example, air and water are not needed for equipment that does not exist. Investment dollars can be saved with fewer fork trucks. Overall, the investment for a lean greenfield site is less than the mass-production plant.

Other Design Issues

The first of several important items to be considered when designing a lean greenfield site is plant layout. Cellular layouts provide the most efficient way to move material through a series of operations. In a cellular layout, each operation is located next to the preceding and following operations. The least amount of distance between operations should be the rule. Short conveyors eliminate the need for material handling. Operators should be positioned on the inside of a U-shaped cell to maximize flexibility when Takt time is changed. A U-shaped layout improves communication between work-team members and provides fast feedback on quality issues. Line-of-sight should be built into the process so inventory and machines do not obstruct operators' views. Since most people are right-handed, material should flow from right to left, or counter-clockwise, in the cell.

One-piece Flow

One-piece flow reduces work-in-process and results in better quality. Work-in-process requires storage (waste), creates repairs (waste), and requires material handling (waste). Quality is improved because there is nearly immediate feedback to the preceding operations.

Triggers

Pull systems minimize inventory and react better to changing demand. They replace only the material that has been consumed (refer to Figure 18-2). Triggers (Kanbans) send signals to the preceding operations to make what has been used. No extra inventory is stored based on anticipated future requirements. Customers change their minds, and pull systems react to the changes by signaling only what has been consumed. Preceding operations are not producing to a forecast, but to actual need.

Material Delivery Systems

Marketplaces and material delivery systems must be established based on planned inventories for every part. In the leanest of environments, marketplaces are used only to store heavy parts that must be delivered by forklifts. The material handler is signaled by the operator (Kanban or Andon) that replenishment is needed. Material within the proper weight limit is delivered directly to the line-side location. Most manufacturing facilities do not have parts delivered from their suppliers every two hours. Therefore, multiple marketplaces are required.

Marketplace Locations

General marketplace locations should be based on the material handler's ability to cover the defined area (usually more than one department). Minimum and maximum container quantities are identified at the general marketplace location. Part quantities are not used because it is much easier to track containers. Operator triggers identify the delivery needs of heavy or oversized parts, called "call parts." Area marketplaces store lightweight containers for delivery to the line side. Minimum and maximum container quantities are identified at the area marketplace location.

Container Quantity

Container quantity is standardized based on usage and ergonomic considerations. Defining material-handling routes will dictate where and how often the material handler travels the route to pick up Kanbans and deliver containers to line-side locations. Line-side inventories are stored at each person's station, and minimum and maximum container quantities are identified at the line-side location. Flow racks should always be used to store card parts at the line side for first-in, first-out control.

Material handlers driving their routes pick up Kanban cards in mailboxes, which signals a replenishment request. On the next pass, the handler delivers the signal-requested product, and again picks up the cards in the mailbox. This process is followed continuously.

Operator Control Systems

Operator control systems prevent defects from moving to the next operation. Andon systems are used by the operator to signal the need for other resources. A centrally placed Andon board notifies the team leader, supervisor, maintenance, material handling, and quality control, etc., that help is needed when an operator pushes the button. The practice of Jidoka (quality in the process) means the operator does not pass a defect to the next operation. Andon systems help create a quality environment.

Customer Demand

The plant's layout should be designed to support expected customer demand. Takt time is used to express the customer demand rate in seconds. (For a complete discussion on Takt time, see Chapter 1.) If the Takt time is greater than the cycle time, capacity is available. If the opposite is true, overtime or more processes are needed to address the required capacity. The objective is to have a cycle time just below the Takt time requirement to meet demand but minimize resources and capital. Meeting the requirement results in right-sized equipment and a smaller plant footprint.

Meeting Places

There are some secondary considerations for planners to keep in mind, such as creating a designated place in each work area for the team to meet. These areas should be large enough for a communications board (usually 4 × 8 ft), individual lockers, a picnic table, and a water fountain.

Aisle Width

Aisle widths should be minimized because specific material-handling routes result in less aisle traffic. Wherever possible, aisles for employees should be designed separately from material-handling aisles. This reduces the risk of injury. If an operation has assembly areas where no call parts are needed, the aisle width can be sized to fit the route of the delivery tugger. Since no fork trucks will be turning into assembly line-side storage areas, the needed aisle width is reduced to roughly the width of the tugger. High bay storage, as well as automated storage and retrieval systems, are not required because of the minimal work-in-process needed to run a pull system.

Separate Manpower/Machine Power

When designing cell layouts in machining areas, it is important to separate the man from the machine. Because of varying equipment cycle times, it is likely that operators can run more than one piece of equipment. To run equipment simultaneously, error proofing must be built into the processes so that the equipment stops when an abnormality exists. Error-proofed processes prohibit defects from moving to the next operation by preventing the error from occurring. If an error is detected, error-proofing mechanisms stop the process. (For a complete discussion on error proofing, refer to Chapter 12.)

Using Robotics

Robotics may be used in place of a person when safety dictates that it is dangerous for a human to perform that operation, or when quality requirements cannot be met by manual operations. A cost analysis of labor versus robotics will usually lean toward robotics as the lower-cost alternative. True lean manufacturing environments do not use robotics to replace labor. With a right-sized labor force, the freed-up labor is used to work on Kaizen (continuous improvement) processes.

Equipment Suppliers

Equipment suppliers are a critical resource during the planning phase of a greenfield site lean implementation. Past equipment may have been over-designed, operated with cycle times less than expected Takt times, or generated more force than lean equipment requires. Right-sizing equipment is critical at this point in the process. If Takt time requirements indicate that a single assembly line needs to produce 120 pieces per hour (30-second Takt time), all of the supplying equipment should make parts at the same rate. The setup goal should be a changeover within cycle time, although initially it probably will not be attainable in all operations. However, as equipment/tooling suppliers learn more about lean manufacturing, setup times will become closer to the ideal of setup times that equal cycle times.

If an expensive piece of equipment must be separated from the one-piece flow process because of long setup times, it should be required to run every part, every day, to minimize inventory and maintain the proper focus on changeover/setup reduction. Changeover time will be at a premium. Tooling and equipment vendors must design their products to handle quick changeovers. Make sure vendors are right-sizing ordered equipment and tooling, and building them to handle rapid changeovers. One-piece flow stops when changeovers are not within cycle time.

Equipment and tooling suppliers can have a large impact on investment and operating expenses. As an example of this impact, assume that a facility's Takt time requirements indicate the need for a medium-speed 100-ton press. Traditional mass-manufacturing analysis suggests buying a high-speed 100-ton press. Business case logic for lean differs in a number of ways, including choosing the right-sized press. Make sure the company uses lean business case logic focusing on total cost. (For a full discussion on the lean business case, refer to Chapter 5.)

Figure 18-3 illustrates another example of potential impact. In this situation, tooling suppliers should be directed to provide quick changeover technology for the new tools. In the mass manufacturing example on the left side of the figure, only two parts are produced. In the lean plant example on the right, eight parts are produced in the same amount of time.

Items that should be required from equipment manufacturers to support lean principles include:

- maintenance port accessibility while the machine is operating (reducing downtime and allowing operators to assist in the process);

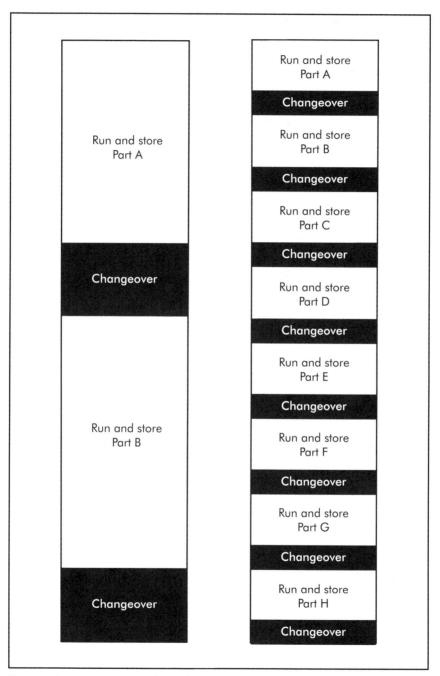

Figure 18-3. Mass versus lean changeover efficiency.

- equipment designers using the "lube for life" strategy;
- color-coding of maintenance ports that provide easy location;
- color-coding of internal piping for air, water, and oil;
- equipment gages that have normal operating ranges visually identified, and
- arrows indicating direction of flow for materials, as well as air, water, and oil, etc.

In addition, error proofing is an important factor for equipment suppliers to consider. Equipment design engineers should require that error-proofing methods be included in each purchase. Prevention is always the first choice in error proofing.

Tooling suppliers must support quick changeover capability. Asymmetrical locator pins, quick disconnects, and clamps versus screws are a few strategies. Suppliers should also minimize the number of idle stations in tooling. Remember, any station in which work is not performed on the part is considered waste and should be eliminated (or not designed into the equipment). Color-coded tools should be provided to make identification easy. (Chapter 14 discusses quick changeover in detail.)

CONSTRUCTION

Construction of the greenfield site should focus on building the facility, equipment, and tools to lean specifications, along with screening, hiring, and training the new workforce. The engineering group must work closely with all contractors to ensure that project timelines are met. Human resources personnel should focus on staffing the new site/project with the best people available at the correct times as identified in the project plan.

Lean Facility Considerations

First, planners must make sure that lean layout specifications are followed. Specifications should include color-coding of all piping and conduit, and flexible conduit and piping for equipment drops where possible—in a lean environment, continuous improvement requires that equipment can be relocated.

Lean Equipment Considerations

Right-sized equipment specifications are crucial when determining the capital investment required for the project. Do not buy a Rolls Royce when a bicycle will meet transportation needs. The equipment purchased should be ready for a lean environment. It should have easy-access maintenance ports that enable operators and maintenance personnel to save time during the process.

The equipment should include a "lube for life" strategy. Self-lubricating bearings reduce the need for mechanical downtime. All lines coming into the equipment should be color-coded for easy identification. Time is critical in a lean environment. Color-coding reduces maintenance time. Equipment should be designed for live maintenance when possible. Downtime is waste, so maintaining equipment during uptimes reduces waste. Sensor systems should be included in the equipment and tooling designs. Sensors prevent machine/die wreckage and prohibit the operation from making a defective part.

Lean Tooling Considerations

Right-sized tooling, with idle stations eliminated or minimized, reduces waste in the process. Quick changeover strategies designed into the tooling reduce downtime associated with setup and changeover. Make sure the production departments sign off on die capabilities.

Lean Work Force Considerations

A lean work force is virtually the opposite of a mass work force. Consider the following characteristics of a lean work force. Operators in a lean work force:

- are part of a team that makes decisions on the working environment and processes;
- do not depend on the supervisor to communicate the next job;
- are totally responsible for the quality of the parts produced;
- are trained in all of the team's jobs;
- focus on continuous improvement;
- signal when help is needed;

- control the amount of component inventory via a Kanban system;
- have a place for everything and has everything in its place;
- have been trained extensively on the requirements for working in a lean environment;
- are part of a team that makes decisions on the working environment and process, and
- help other team members.

TRAINING

It is critical when establishing a continuous improvement philosophy at the new site to train operators to work in a lean environment. Operator training must be provided in a number of areas, beginning with the lean philosophy overview. The initial training session should identify the goal of lean manufacturing—the elimination of waste. Specific training topics should be addressed briefly in the overview presentation.

Topics

Teamwork in a lean environment espouses that each person is a member of a work team. The importance of teamwork should be addressed regarding problem resolution, work-group skill requirements, quality improvements, workplace organization, etc.

Conflict resolution is important because the dynamics of teamwork are such that conflict occasionally develops in work teams. Understanding the use of consensus decision-making to resolve conflicts results in unified work teams.

Problem solving (the 5-Why analysis) is critical in a lean environment because each member of the team is expected to contribute to problem resolution. Using a 5-Why analysis approach gets to the root cause of a problem. Effective countermeasures eliminate the problem and prevent its recurrence.

Workplace organization is a core characteristic of lean enterprises. A basic tenet of lean manufacturing environments is that production and nonproduction items are well-organized and every required item is placed in a specific location. Employees need to understand this concept and its application in the production of quality products.

Pull systems are only effective when operations are stable and predictable. Employees have to know that warehouses in the new facility are not to be filled with inventory. Inventory levels at each workstation only support a few hours of production. A Kanban (pull) system is used to signal when more components are needed. Operators must understand how to establish minimum and maximum levels of inventory in their workstation, as well as how to trigger replenishment.

Standardized work instruction is the basis for quality production because each person performs the same job in the same way. Standardized work training accomplishes the "how to" of:

- detailing each work step;
- graphing the time required for each step;
- defining safety equipment required for each step of the process;
- defining quality checks, and
- visually representing the workstation layout, with a step chart showing the operator's walking path to get each component and perform each operation.

Error proofing builds in quality at the process. In a lean environment, each operator is responsible for his or her production quality. Error-proofing devices built into the operation reduce the number of defects produced. Operators must understand how to conduct a Pareto analysis of defects, then apply error-proofing countermeasures to eliminate the possibility of producing them again.

Quick changeover is the lean manufacturing method for reducing setup times. Most operations require changeover from producing one component to another. Employees must know the basics of quick changeover to reduce downtime, and be aware of a number of factors to reduce setup times, such as:

- separating internal from external elements;
- tool boards for locating changeover tools;
- timing of resource availability, and
- quick changeover tooling methods.

Routine maintenance procedures to be performed by the operator must be defined. In a lean environment, a number of maintenance tasks now performed by skilled tradespeople should be done by operators. Checking fluid levels, monitoring gages, keeping equipment clean, and using the basic senses (touch, hear, see, smell) to determine whether the machine is operating properly are a few of the tasks that operators can assume.

A *visual factory* is an environment in which anyone can walk into an area and understand operational flows and performance issues within 10 minutes. The new facility should focus heavily on visual factory techniques, such as:

- floor markings for dedicated locations;
- communication boards with updated graphs of information on performance items;
- color coding;
- gages identified with normal operating ranges, and
- inventory minimums and maximums.

Operators must be aware of each item's importance.

Quality in the workstation is one of the keys to lean manufacturing success. Lean companies focus on producing quality parts. Repairs and scrap are waste. Every operator must understand that he or she is responsible for sending only good parts to the next operation. Fixing the defect in a repair station down the line is not possible.

Job rotation is required to ensure that each person knows how to effectively perform every operation in the cell. Each team member is responsible for knowing all of the team's jobs. Most lean environments rotate operator positions every two hours when the job requires using different muscle groups. Rotating each position reduces the possibility of carpal tunnel or thoracic outlet problems, and ensures well-trained crews.

Andon systems are a quick means of identifying that an operator needs assistance. Andon boards use signal lights to let the team leader, maintenance, or material replenishment people know their help is needed at a specific operation. Employees have to know how Andon systems work.

Continuous improvement (Kaizen) drives lean enterprises to constantly evaluate all processes while searching for better ways to perform them. In training sessions, the initial workstation layouts and production processes should reflect the best method known to effectively make products. Everyone is expected to help the company reduce waste by identifying better methods for producing parts, reducing defects, improving space utilization, etc.

Training for the *physical jobs* of the team takes place prior to startup. A timing chart of all operators in the cell should list the expected time required for each person to learn every job. The progress of each person's ability to produce quality parts within Takt time should be monitored, and additional training should be provided as needed to bring each person up to speed. The amount of pre-startup training depends on the complexity of jobs. Allow appropriate training time for employees to learn how to perform each operation. Many companies have started plants with insufficient training, then suffered high overtime costs because of poor productivity and low quality.

Successful companies focus on *total cost*. Most managers are familiar with budgeting in a traditional mass environment. In a lean facility, the focus on operational and financial issues moves away from individual budgets toward total cost. It should be made clear that the operation's total cost will be used to drive decision-making. In some cases, an improvement in one area (A) can cause problems in another area (B). A budget approach allows changes in area A, resulting in problems for area B. Management then uses a bigger stick to try to force area B back in line. The total-cost approach would deny the change in area A, recognizing the impact on the bottom line.

Managers also need some training in the areas shown above. It is important that each manager understands the job requirements in his or her area. It is equally important that each manager have a sound understanding of lean philosophies and how the company is applying them so full support can be given.

Obviously, the training process will be time-consuming and require significant expense. In the long-run, however, training expenses are returned to the company many times over by the resulting lean operation and established continuous improvement process.

STARTUP

Construction is complete, machines are in place, training is finished, and the work force is ready to go. It is time to begin making products. During the startup phase, managers' focus is on learning to build products and making sure support systems are in place while the production pace is still slow. It is critically important for lean philosophies and practices to be maintained. As problems arise, some may want to revert to mass thinking. Eliminate this thinking as soon as possible. Items to focus on during the startup phase range from establishing team concepts to providing communications boards for the teams.

Lean Metrics

As mentioned earlier, lean metrics are part of an operational philosophy that distinguishes lean manufacturing from mass manufacturing. Operational metrics focus on overall equipment effectiveness, dock-to-dock and first-time-through methodologies, building to schedule, and total cost. (For a full discussion of lean metrics, refer to Chapter 3.)

Establish the Team Concept

Each operator is part of a team. The size of the team is mandated by the physical area available. Teams should be limited to six people, including the leader. Small teams are better able to communicate with each other, so they will address issues more quickly and effectively. In many cases, however, teams are larger than the ideal of six people. Figure 18-4 shows the most efficient organizational layout based on 10 team members.

Training for Primary Operations

Operators must grasp their own process first, before moving on to the other jobs of the team. As their skill level reaches competence, they can be moved to other jobs within the team. The chart in Figure 18-5 is a good way to track skill attainment. By looking

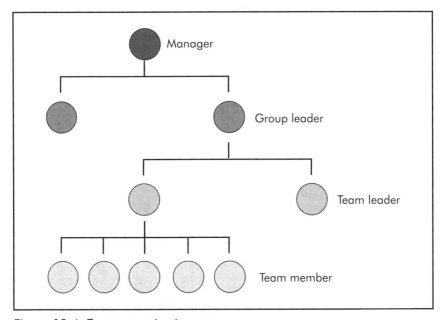

Figure 18-4. Team organization.

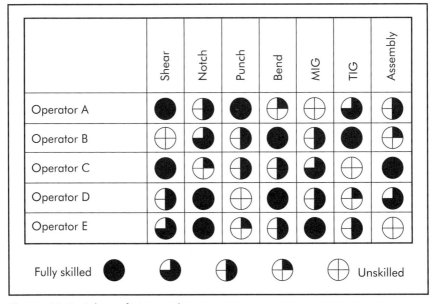

Figure 18-5. Job proficiency chart.

at the chart, it is easy to determine the amount of training every operator has had on each of the team's jobs.

Standardized Work Instructions

While operators are learning their new jobs, they should begin working on standardized work instructions (SWI). The first step is to identify each element of the job. The example in Figure 18-6 has identified five wasteful processes (only the fourth step adds value), along with the amount of time each process takes. The graph for time spent in wasteful operations can even be color-coded to separate value-added from non-value-added operations.

Identify a step chart on the SWI that shows element locations. Figure 18-7 is an example of a step chart that shows wasted motions in the process. It clearly identifies that some workplace organization is needed—moving storage locations closer to the machine and reducing the time spent walking.

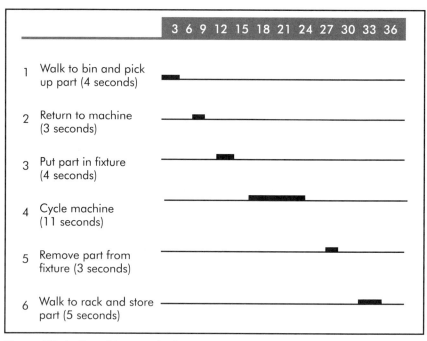

Figure 18-6. Graphing work elements.

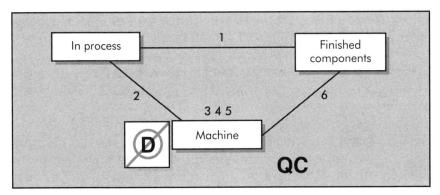

Figure 18-7. Step chart.

Illuminate any safety requirements, such as hearing and sight protection, steel-toed or metatarsal shoes, gloves, etc., on the SWI. In addition, the manager must identify lean tool implementation, such as error-proofing and quick changeover, shown in Figure 18-7 as ⌀ and QC.

The final area in SWI is approvals. Each operator must sign off that the information is correct. Other signatures may include the team leader, group leader, and area manager. Safety and ergonomics managers should also sign off, as well as an engineering representative.

As operators learn their new jobs, developing the SWI becomes a useful learning tool in that it:

- reinforces sequential work steps;
- identifies areas of waste, which gets people thinking about improvements;
- specifies safety equipment required for the job, and
- becomes a training tool for new operators.

If multiple shifts are in operation during the startup phase, it is critical that operators from each shift agree on everything listed on the SWI. Each person must do the job exactly the same way so that variability is reduced and quality is a given.

Team Communications Board

The team communications board becomes a working tool during the startup phase. The communications board houses data and information important to the team, work group, or department. Data is usually reported for safety, quality, production, cost, and morale, along with other categories, such as maintenance activities. Figure 18-8 shows an example of charts that may be found on a team communications board. It is best to make the board portable.

The startup phase also begins to validate internal and external logistics systems. Problems that appear during startup must be effectively countermeasured (eliminated) at this time. Failure to do so results in significant problems during ramp-up and normal

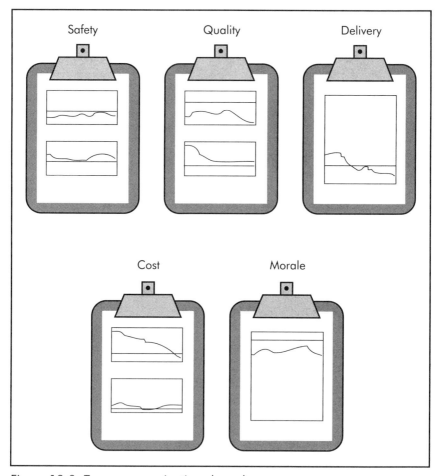

Figure 18-8. Team communications board.

operation phases. The last step during startup is to finish plans for the ramp-up.

RAMP-UP

During ramp-up, production rates increase at a scheduled frequency. As rates increase, all systems are tested. As problems develop, modifications may be necessary to smooth them out. Some people may want to implement mass-production countermeasures such as, "let's add some inventory at operation 40 so we don't run out." The trick here is to stay within the tenets of lean manufacturing. Adding inventory is not the answer. First, determine the problem with data, then employ countermeasures that focus on lean solutions. The following sections discuss expected problems during ramp-up.

A Line Appears Imbalanced

The line must not be changed until operators' skill levels have been evaluated (training charts will provide the information). Review the SWI at each operation and verify that each operator is adhering to them. Is there a more effective way to organize the workplace? Does the material presentation ensure minimum waste when moving material within the operation? Have work elements changed for some reason? The key is to investigate the situation and use lean tools to countermeasure the problem.

Excessive Downtime

If implementers encounter excessive downtime, verify that the operator is cycling the equipment as specified, and that preventive maintenance is being performed as required. A downtime Pareto analysis can be used for clues on where to apply countermeasures.

Operators Wait for Components

If there is a problem in an upstream production operation, or the material delivery system, upstream operations may need to be reviewed for machine/operator availability or downtime problems. If the issue is material delivery, material handling routes should be reviewed to verify that delivery schedules can be met. Material handlers must communicate their problems. A scheduling system could be the problem. Lean environments use a pull system of Kanbans. Material is produced only when it is needed, not when a scheduling system says it should be produced.

Team Meetings are Cancelled

Team meetings show the company's dedication to lean manufacturing. Canceling meetings sends a negative signal to the teams. Overtime should be scheduled to meet the backlogged production in lieu of canceling planned meetings.

Quality Problems

Speeding up the machining/assembly process may cause quality problems to surface. In many cases, standardized work is not followed because of time pressures. In other cases, error-proofing opportunities become apparent. Regardless of the reason, use root-cause analysis and develop a countermeasure. The plan-do-check-act (PDCA) cycle is the most effective way to solve problems.

The key to solving ramp-up problems is using lean techniques. Mass-production countermeasures dilute a lean site by introducing waste into the operations.

NORMAL CONDITIONS

If the previous processes have been followed effectively, the normal production phase focuses on Kaizen processes. If not, time and resources will be spent correcting problems that surfaced during the startup and ramp-up phases.

Operating teams are the primary drivers of the Kaizen process. They use data to identify problems and develop countermeasures. The implementer's job is to provide support for their continuous improvement efforts.

CONCLUSION

Making changes in an existing brownfield site requires working around monuments and existing practices. A greenfield site implementation is much easier—it enables a company to start with a clean sheet of paper. As that paper begins to represent the new greenfield facility, the layout should take advantage of the lean principles and techniques identified in this chapter. The operational result will be a new facility with:

- a smaller footprint requiring less investment;
- less dedicated inventory space;
- cellular operations using one-piece flow;
- team-enhanced operations;
- quick throughput times;
- improved quality;
- exceptional workplace organization;
- lean metrics, and
- better profitability.

Brownfield Site Implementation 19

By Richard Dixon, Dennis Raymer, and David Stewart

Most readers may be considering or are already engaged in a lean conversion of an existing facility. As noted in earlier chapters, a *brownfield implementation* is one that takes place in an existing facility, while a greenfield implementation occurs in a new facility. Although many may wish to design a new plant from scratch that incorporates the best of lean manufacturing, the reality is that few have the luxury. This chapter focuses on some of the issues involved in brownfield lean conversions.

Many industry leaders have recognized the value of lean systems and are asking their employees to begin making the changes that will turn their plants into world-class lean facilities. There is hope that such transitions will be seamless, with little impact on production or morale. Realistically, considerable chaos accompanies brownfield conversions despite careful planning. Making a transition this large in scope is inherently disruptive, and people typically have a hard time dealing with changes of this magnitude. Given the difference between expectations and results, it is important to be aware of and plan for a phased implementation that introduces different elements and changes at different times.

Earlier chapters have discussed the five-phase model of implementation (Chapter 7) and change management issues (Chapter 6). This chapter discusses how these items apply in brownfield settings after introducing the reader to some of the ways that brownfield and greenfield implementations differ.

EMPHASIS ON CHANGE MANAGEMENT

Brownfield conversions are considered much more difficult to implement than greenfield startups. Rather than starting with a blank sheet, brownfield plant managers must concern themselves with how to make significant changes in virtually every aspect of their operation. Often, significant transitions of time and effort are required, particularly in plants where the work force has been in place for many years and a strong culture exists. If no recent history of successful change initiatives exists, the task becomes even more difficult. A lean implementer can expect considerable resistance to any proposed changes unless the conversion is managed exceptionally well. Attention to the principles of change management is a strong component in successful conversions. Lean implementers must not expect employees to embrace all of the changes without thoughtful consideration of the demands placed on all.

Transformation of Culture

The approach to lean in an existing brownfield facility is quite different from that in a facility started from scratch. In a new facility, there is no existing work force accustomed to a certain way of doing things. This is an issue in brownfield plants, particularly when the work force has been in place for 10–20 years. A standing plant has an existing culture that has developed over the years and a lean initiative will challenge that culture and, if successful, replace traditional mass-production values and behaviors with those compatible with lean. Cultural change is not a simple process. To change a culture often means a commitment measured in years.

Changes in Habit and Behavior

Employees comfortable with a certain way of doing things that worked in the old culture will need to learn new behaviors and undo old habits to succeed in the new lean culture. Since the production system dictates the new culture (see Chapter 8), certain behaviors will work in the new culture and others will not. Old ways of thinking (traditional mindsets) must be replaced by new lean mindsets.

Conversion of Systems and Procedures

In a greenfield site, systems and procedures can be designed to make lean work in the plant before the first brick is laid. In a brownfield setting, the emphasis is on conversion from old systems and procedures to new ones. Old layouts and job assignments must be changed, and employees must adapt to new setups and different work steps while maintaining production. Teamwork must be learned by operators in settings where the prevailing mood has been "we're all in this alone." Skilled tradespeople must adapt a new emphasis on prevention, rather than firefighting. Material handlers are used less, and some may even be redeployed, while others learn new patterns and job expectations.

THE LEAN CONVERSION OF BROWNFIELD, INC.

The best way to illustrate what is involved in a brownfield conversion is through the following case study, which has been compiled from a number of different implementations. Any resemblance to an existing plant or company is purely accidental. This study will show those charged with implementing lean in their facilities:

- how to approach the process;
- the importance of a preliminary assessment and planning period;
- the value of a phased implementation, and
- the importance of using metrics that provide an ultimate goal and a measure of progress.

This brownfield conversion is typical of plants still engaged in traditional mass-production practices. This particular operation has several plants in the Midwest. The plant this study focuses on has been in business for more than 20 years. It averages a dependable $160M in sales every year due to major contracts with longtime customers. Recently, however, plant leaders have become concerned that their share of the market will begin to slip as competition emerges. Up until now, they were the only game in town.

There are 600 employees in the plant, many of whom have worked at the plant since it started. The relationship between the workers and management has been good, albeit somewhat paternalistic and patronizing. Expectations have not been set too high and the workload has not been overbearing, so the employees think it is a good place to work.

Decision to Implement Lean

Concern about competition has plant leaders thinking about a lean conversion. Total Systems Development, Inc. (TSD) was referred as a company to provide them with consulting and training in the conversion process. The company's goals included improving quality, improving customer service and delivery, improving Just-in-Time (JIT) delivery within the manufacturing process, and lowering the cost of operations.

The plant's implementation process began in March and was ongoing through the next spring. In this period, the client saw the typical benefits of a lean implementation. Inventory was decreased from $680M to $210M. Productivity per direct laborer increased from 1.38 units/person/day to 2.09 units/person/day. Required assembly space was reduced by 30%.

Assessment and Mapping

After agreeing on the goals, the TSD consultant/project manager suggested as a next step a value stream mapping and assessment to provide data for planning the implementation. (For a complete discussion on value stream mapping, refer to Chapter 4.) TSD's mapping and assessment is a five-step process that involves:

- developing a current state map of the targeted areas,
- TSD's lean simulation, Charlie's Coil Factory™,
- an assessment using TSD's Lean Diagnostic Tool™ (LDT),
- TSD's Calibrate™, a lean diagnostic assessment tool,

- the creation of the future state as defined by Calibrate and mapping, and
- developing the business case for lean implementation.

Calibrate is an assessment of 42 lean elements organized by eight components of a lean production system. This assessment provides data that serves as the baseline for measuring progress, and provides a means to organize the work plan. The TSD project manager stated that, from the data gathered during the assessment process, a business case for the implementation would be developed and a compelling return on investment demonstrated.

The mapping process recommended by the consultant and chosen by the client involved a mentoring program at each step of the implementation. The client selected a manufacturing supervisor and an assembly supervisor to receive the mapping knowledge transfer. They, in turn, were expected to conduct future mapping in other areas of the plant. The mapping process commenced with the supervisors and project manager visiting the primary assembly area. They mapped each of the processes and collected data along the way. As elements of waste presented themselves, short, single-point lessons were used to identify the problems and start discussions about possible improvements. On many occasions, the manufacturing supervisor said, "we can't do that here." After a few days, the recommended phrase, "how can we do that here?" became the standard. It was a welcome relief to see the mass-manufacturing wall around that supervisor beginning to come down.

It is easy to understand people's resistance to change. At the client's site, most of the employees only knew one way to produce their products—plenty of inventory, overtime, and rejects. Their world was about to change. The decision was made to introduce lean to the plant one department at a time. For the mapping and assessment of the facility, one assembly line was chosen as an initial application area. The results at the end of this case study apply to this first effort to introduce lean to a specific area within a mass-production plant.

Charlie's Coil Factory

While the mapping process was proceeding, the consultant presented his lean simulation, Charlie's Coil Factory, to the plant's management staff. The simulation is an effective tool to help people get an experiential feel for the change process involved in a lean implementation, as well as realize the striking difference in lean metrics contrasted with mass metrics for measuring the cost of manufacturing. Charlie's Coil Factory consists of four rounds of simulated production activities, each round representing different levels of lean implementation. Upon completion of the last round of the simulation, the plant's management group agreed that lean certainly produced better results than the mass approach. They were quick to point out, however, that the simulated plant's situation in round one was much worse than their current operations. That fact was true, but it was also true that each of the wastes represented in the simulation existed in the assembly department chosen for the initial lean conversion. The value stream mapping proved that statement correct.

Calibrate—Current State

As the current state mapping was nearing completion, the next step of the assessment process was initiated. Calibrate was used to assess the gap between the client's current state and the desired future state. The radar chart in Figure 19-1 indicates the client's lean readiness at the beginning of the implementation. World-class performance is shown as Level 5 on the graph.

Scores for the current state in Figure 19-1 are typical for a plant that is ready to begin lean implementation. Higher scores are shown in the first two categories of leadership vision and commitment and change management, and reflect the fact that some preparation, education, and discussion has already occurred to create receptivity to lean implementation. The lower scores in the next two categories were also expected. The plant's organizational structure and support system were created with mass-production goals in mind; work must be done so the structure and systems support lean practices. The corporate culture and workplace climate also must be changed to reflect lean values and principles. Low scores are expected in the technical areas of equipment stability, in-station quality, and just-in-time production, since there has been no effort to make changes on the shop floor. Finally, continuous improvement is a goal of the lean implementation and typically is the last category to improve.

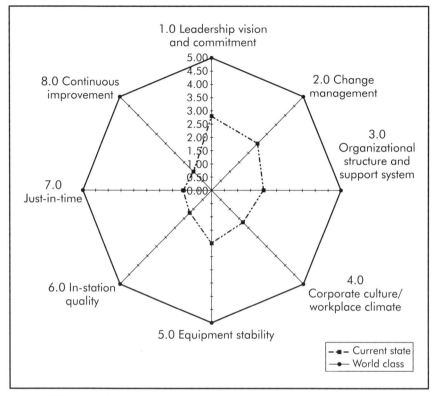

Figure 19-1. Calibrate—current state.

The Current State Map

Company managers and the consultant reviewed the results gathered from mapping the current state (shown in Figure 19-2). The forecast of customer requirements was 65 units per day for this single-shift department. Although a power-and-free conveyor was used on the main assembly line, inventory was allowed to build up between stations. Without a Total Productive Maintenance (TPM) program, the conveyor was having continuous clutch problems that resulted in an estimated 95% uptime rate. Downtime was not tracked. Cycle times varied significantly between operations on the line, as well as between the line and subassembly operations. The first operation determined the sequence down the line—no changeovers were required. A significant amount of overtime was being worked to meet the customer

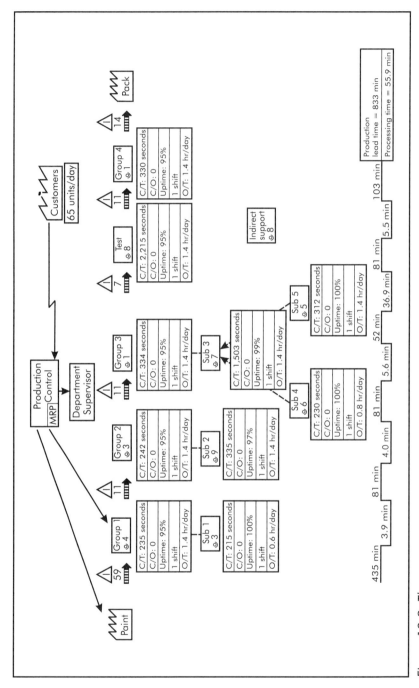

Figure 19-2. The current state map.

demand. Value-added assembly time accounted for 6.7% of the total throughput time.

The following details provide a better understanding of the issues with which the consultant and the client were dealing.

- The line was set up with one-piece flow but was not paced, even though a power-and-free conveyor system was in use on the assembly line.
- Operators on the line and in subassembly areas worked at a slower pace than that required by the customers. With cycle times higher than Takt time, overtime was required to meet demand.
- Large containers of materials were stored at various locations along the lines, sometimes representing weeks of material.
- The line was running out of fabricated materials frequently, but would push the unfinished assemblies to a back line and continue working on assemblies for which material was available.
- The quality of purchased and fabricated parts was a daily problem.
- In-process material stacked up as operators periodically left their workstations.

Calibrate—The Future State

The next step in the process involved developing a future state map for the client's primary assembly department, which accounted for roughly 40% of the daily shipping volume. Calibrate had identified management's vision of the future, so the job was to put that vision on paper with respect to the primary assembly operation. Figure 19-3 shows the client's future state vision, compared with world-class standards and the client's own current state results.

The client chose levels slightly below world class for this implementation. Budgetary and time constraints dictated the target level. Emotions ran high during development of the future state. The implementation team engineer guaranteed that assembly and the supplying subassembly operations were balanced and manned correctly. They were, when working within the constraints of the

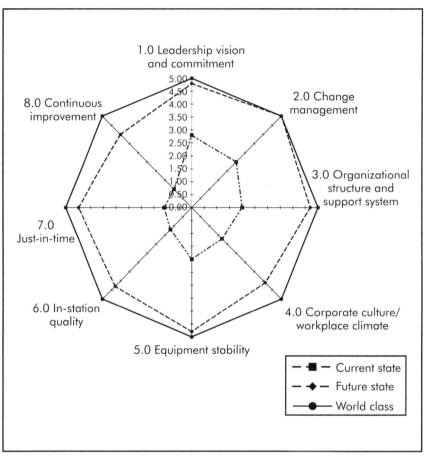

Figure 19-3. Calibrate—future state.

department's layout and the engineer's knowledge. But the consultant was looking at the process with "lean vision." The wastes were easily visible, as were the lean methods to reduce them.

Future State Map

Review the improvements addressed by the future state map in Figure 19-4. With the customer demand target increasing to 75 units per shift, Takt time was calculated to be six minutes, or 360 seconds. Some of the other improvements identified are as follows:

- A marketplace was established between the assembly operation and the paint department. Pull system Kanbans were used to trigger replenishment from the paint department.
- Production control used a heijunka (level scheduling) process for mixed-model scheduling of the line.
- Organization of each workplace included flow racks with minimums and maximums, along with returnable containers for 90% of the component parts. Material was replenished nightly.
- Synchronous production was accomplished using a scanning program that electronically fed information to the subassembly areas. When the first operator loaded a unit on the conveyor, he or she scanned the model, which broadcast the model requirements to the downstream operations.
- By rebalancing the operations to a cycle time close to Takt time, 16 operators were removed from the process and most were used for a new product launch—the rest filled other slots.
- Lean operations required additional team leaders, route delivery people, and a line quality person for a total addition of six people—a net headcount reduction of 10 people for the assembly operations.
- The power-and-free conveyor was replaced with a roller conveyor, eliminating the clutch problems.
- The back line (repair line) was removed from the testing area, forcing quality problems to be addressed as they occurred.
- The fact that fabrication quantities for this assembly line were reduced from four weeks' worth of requirements to one weeks' worth is not visible in the map shown in Figure 19-4.
- The actual value-added assembly time percentage increased to 16.9% of the throughput time, an improvement of 252%.

Creating the Business Case

The final step of the mapping and assessment process was to complete the business case (see Chapter 5). During this phase, the lean consultant's knowledge was critical in determining the potential savings resulting from the lean implementation. Values were estimated using the consultant's prior experience and knowledge.

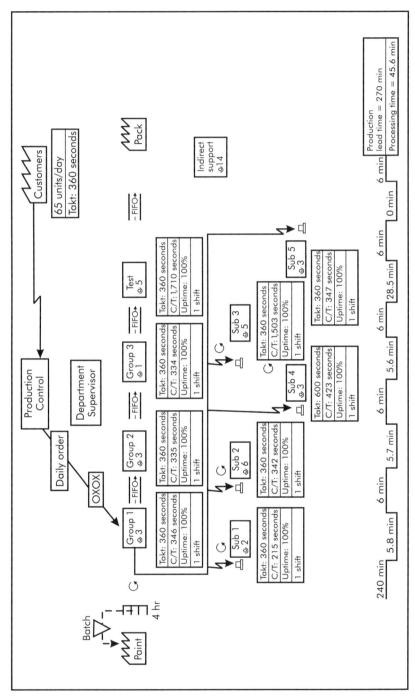

Figure 19-4. The future state map.

Table 19-1. Annual repeatable savings

Direct labor	$683,904
Inventory reduction of $200,000 carrying charge at 20%	$40,000
Added group leaders, material handlers, quality	($269,755)
Total expected annual savings	$454,149

Annual repeatable savings are outlined in 19-1. A cost reduction of $25.64 per unit was realized. Other expected benefits included opening up over 7,600 ft^2 (706 m^2) of assembly space, reducing forklift traffic, reducing overtime, and reducing downtime resulting from the elimination of "lost" material. Overtime reduction would prove to generate an additional $190,000 profit annually for this line. As a result of the data from the mapping and assessment process, the client had no problem getting the capital appropriation request approved by corporate headquarters. Management eagerly awaited the expected results.

The Lean Implementation

The lean implementation process began two weeks after the mapping and assessment was completed. The first step in the process was to develop the project implementation plan. In keeping with the TSD approach presented earlier, the recommendation was for a five-phase approach to lean implementation:

- develop stability;
- create continuous flow;
- standardize work;
- create pull systems, and
- develop level production.

The following outline indicates some high-level points from the project plan:

Stability—target date April 20
- Create a team structure in the department.
- Initiate 5S activities in the assembly processes.
- Reduce/eliminate downtime in assembly operations.
- Improve quality from internal and external suppliers.
- Improve workplace organization.

Continuous flow—target date May 13
- Relocate operations to reduce/eliminate buffers.
- Reduce lot sizes.

Standardized work—target date June 3
- Develop standardized work for each operation.
- Synchronize production for subassembly operations.
- Create one-piece flow.
- Make sure the daily schedule is attained.
- Develop mistake-proofing (poka yoke).
- Implement visual management.

Pull systems—target date July 14
- Create a marketplace to enable Kanban scheduling of painted and fabricated parts.
- Develop line-side flow racks with minimums, maximums, and triggers.
- Store all parts on the assembly and subassembly lines.
- Replenish parts based on usage.
- Create central storage areas.
- Define conveyance routes.

Level production—target date August 1
- Use mixed-model scheduling through the heijunka process.
- Produce a repeatable pattern by volume and mix for each day of the monthly production plan.
- Level production is characterized by making a fixed production plan over an extended time, the ability to make every model every day, daily adjustments, and a predictable production process.

Some of the phases overlapped due to the time constraint of five months for the assembly line implementation. Work on improving stability (the first phase) began March 1.

The Stability Phase

The implementation team's first priority was developing a team environment. The line supervisor identified four people that would make the best team leaders. The consultant met with them as a group and defined the responsibilities of the team leader position in a lean manufacturing environment. All four accepted the position, which included a small pay increase. Over the next four weeks, the team leaders received training in lean principles, team dynamics, problem-solving, and work-site training. They then began to form teams in their areas.

5S. This was one of the first items addressed by leaders to their team members. The red-tagging process showed a great deal of unneeded material, supplies, and other items. Floor fans took up needed space, as did chairs for use during line downtime. In the future lean environment, fans would be mounted above the work areas, chairs were to be replaced with picnic tables in team meeting areas, and line downtime would be virtually nonexistent.

Downtime. The second topic attacked in the stability phase was downtime of assembly operations. The primary hindrance to throughput turned out to be inventory shortages. The secondary problem—conveyor downtime—would be addressed during the July shutdown period, when the power-and-free conveyor is to be replaced by a roller conveyor. The implementation team developed a temporary action plan with production controllers to begin physically verifying parts availability. The process began in the middle of March and showed results almost immediately. Throughput time decreased 11% by April 1.

Quality. The third stability topic focused on quality. The consultant met with the director of manufacturing to discuss methods for improving fabrication quality. The consultant also met with the director of purchasing to discuss vendor quality. Because quality data was lacking, team leaders began tracking quality data for their areas. During the next two weeks, team leaders gathered quality information and shared it with directors. With data in hand, the directors began developing quality improvement plans with their respective suppliers. During the last two weeks of the stability phase, three suppliers met with team members and discussed quality problems. The suppliers developed countermeasures that could be quickly implemented. Fabrication quality was addressed by the quality manager, who began a process to inspect the first and last pieces of parts heading for the lean assembly line. Quality improvements were noticeable within two weeks. Quality rejects were down 27% by the end of April.

Workplace organization. The last stability area the implementation team addressed dealt with workplace organization. Some of the pneumatic tools were repositioned for operators to use more easily. Some component materials were moved closer to the operators to reduce walking time. Operators were pleased with the changes, although measurable improvements were not recorded. Lean metrics for stability improvements are shown in Figure 19-5. The focus on stability items resulted in improvements in all metrics.

Assess the results. A Calibrate assessment was conducted at the end of the stability phase. The results are depicted in Figure 19-6. There were modest improvements in the first three categories, but much more significant improvements in categories 4.0 (corporate culture/workplace climate), and 5.0 (equipment stability). This is consistent with focused improvements in the stability phase. The last three categories (in-station quality, Just-in-Time, and continuous improvement) will receive more attention as implementation progresses through the remaining four phases (continuous flow, standardized work, pull systems, and level production).

The Continuous Flow Phase

While continuing to push for stability improvements, the implementation team initiated the continuous flow process for assembly operations. The first priority was to determine which operations could be moved in the short term to facilitate process flow. The first, second, and third subassembly areas already fed directly to the point of use on the assembly line. They were, however, scheduled to be revised to the future state during the July plant shutdown, as were subassembly areas four and five. The physical relocation was still three months away, so the consultant and team leaders decided to change the process flows within the current layouts.

Only subassembly area two had any semblance of flow, but there were small amounts of in-process inventory between operations on the U-shaped conveyor. By working with the team leader and the operators, inventory on the conveyor was reduced and availability at the point of use was improved. The same short-term approach was used with the other subassembly areas, and inventory in those areas was reduced as well.

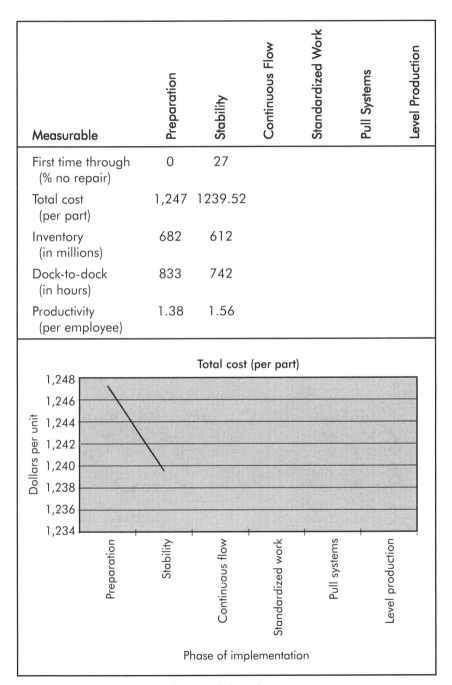

Figure 19-5. Lean metrics of the stability phase.

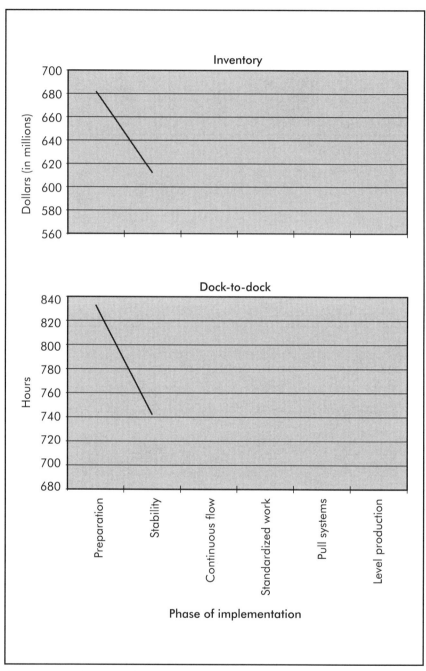

Figure 19-5. (continued).

Brownfield Site Implementation

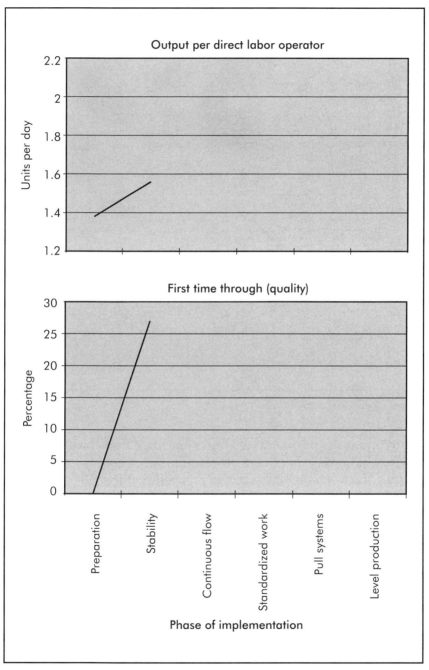

Figure 19-5. (continued).

Lean Manufacturing: A Plant Floor Guide

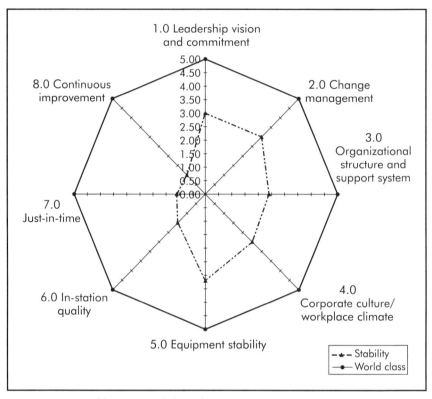

Figure 19-6. Calibrate—stability phase.

During this time, operators began asking about the changes. A departmental meeting about lean manufacturing was held to introduce the team members to the principles of lean. The primary concern of the operators was job security, followed closely by how their jobs would change. Company managers assured them that a lean implementation would have no impact on job security, but that some work tasks would be redistributed in the near future. People were evenly split on whether they liked the changes.

The implementation team continued its focus on continuous flow. The second target in this phase of the implementation was to reduce lot sizes. A returnable container strategy was developed with the purchasing manager, and the director of operations began reducing lot sizes for fabricated parts. There was a short-term availability problem with in-house containers. They were designed to handle mass quantities, but were only being partially filled with lean quantities. The problem was solved when internal, returnable containers began arriving.

Purchasing department personnel negotiated with suppliers to amortize the returnable container costs into the piece-price for the coming year. As suppliers began using the returnables and shipping smaller quantities more frequently, the investment in parts began to drop. An important point to note is that lean is a system, not a collection of tools. One of the fundamental rules in lean manufacturing is to drive decision-making using a total-cost approach. From a purchase price variance standpoint, amortizing container costs into the piece-price drove the price variance up, reducing profit. From a transportation budget standpoint, increased delivery frequency means higher costs. Considering the client's potential was an additional $900M profit per year, it was easy to justify spending $76M on containers and extra transportation costs. In reality, the inventory manager was able to reduce the increased costs by $14M.

Continuous flow improvements were in place by the May 13 target date. The impact on the company's metrics is shown in Figures 19-7. Calibrate was administered at the end of the continuous flow phase. All categories posted gains, with the most significant improvements in categories 5.0 (equipment stability) and 6.0 (in-station quality). The radar chart in Figure 19-8 shows the results of the continuous flow improvements.

The Standardized Work Phase

The priority in the standardized work phase reviewed by the implementation team was to make the daily schedule. The monthly schedule that assembly lines had been working to for years was broken down into daily requirements. Production would flow until the company president received demanding telephone calls from customers or salespeople. Fabrication and assembly would then go into crisis mode to meet the president's request/demand. When a painted model was available, assembly would start the model down the line with the hope that fabrication would get the necessary components by the time they were needed on the line. If not, the model would be moved to the back (repair) line to wait for the

Measurable	Preparation	Stability	Continuous Flow	Standardized Work	Pull Systems	Level Production
First time through (% no repair)	0	27	31			
Total cost (per part)	1,247	1,239.52	1,232.77			
Inventory (in millions)	682	612	566			
Dock-to-dock (in hours)	833	742	706			
Productivity (per employee)	1.38	1.56	1.67			

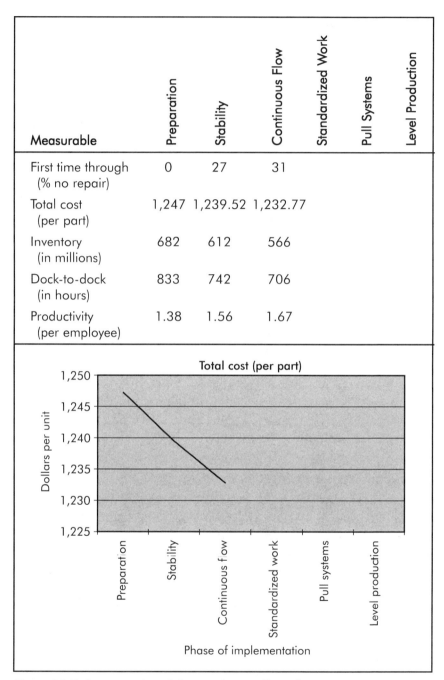

Figure 19-7. Lean metrics of the continuous flow phase.

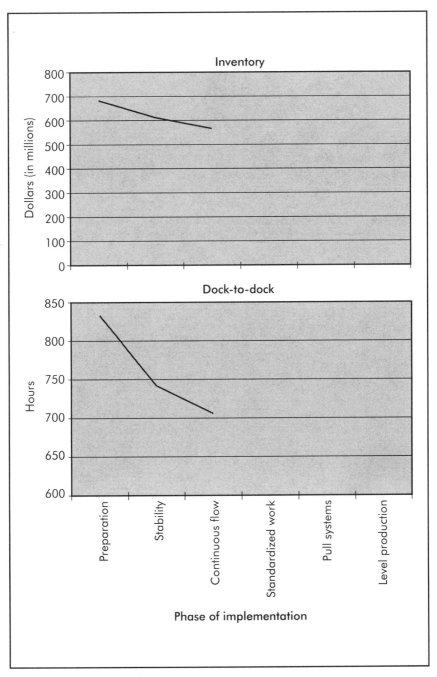

Figure 19-7. (continued).

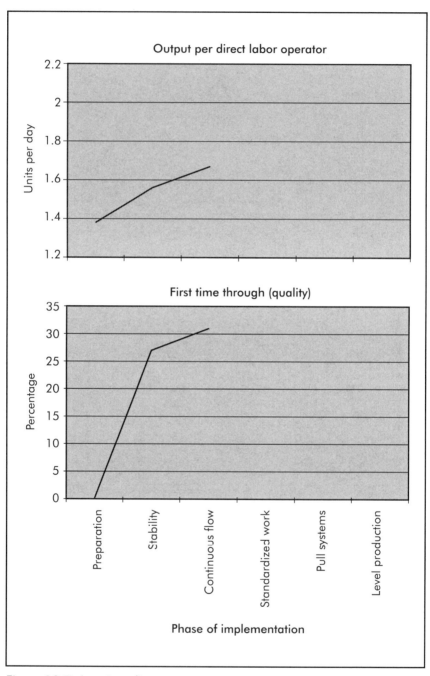

Figure 19-7. (continued).

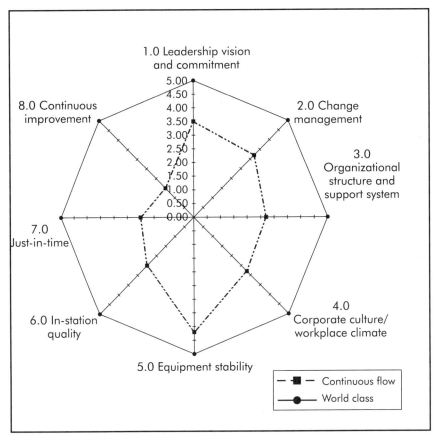

Figure 19-8. Calibrate—continuous flow phase.

components to arrive. The availability of the back line enabled line operators to start units that could not be completed on a timely basis, thus impacting their ability to satisfy daily schedules.

The president's request always had a major impact on fabrication and paint. Each department had to suffer through teardowns and setups to get the needed components. The future state for the assembly line focuses on pull systems and level production, which minimize the impact of telephone calls to the company president and create a high probability of achieving the daily schedule.

Developing a standardized work system. After a daily schedule was achieved, the implementation team reviewed the development of a standardized work system for each operation.

With major layout and work content changes coming in July, it was decided to forego the implementation of standardized work until after the changes. Training was provided to the local members of the implementation team, along with two industrial engineers. The engineers created instructions for a standardized work system (for existing jobs) in the assembly department as a training exercise. The conclusion of the company's implementers was that creating standardized work would be difficult for operators, so engineers should do the work. The consultant opposed this recommendation to management, but was overruled. When operators create standardized work, they take ownership of their processes and in the continuous improvement of their processes. Without that ownership, empowerment is just a word.

Management explained that a tight budget and peaking customer demand would probably result in insufficient funds and inadequate time to allow operators to create the standardized work. By using the word "probably" the door was left slightly open regarding operator involvement in creating standardized work after the rearrangement.

Synchronizing production systems. The implementation team then directed its efforts to synchronizing the production of subassembly operations with the main assembly line. Each of the subassembly operations had buffers between it and the next process. Sub-areas four and five were physically distant from their customers, so some inventory was necessary. It was decided that these two areas should use visual Kanbans as their replenishment tool. As finished subassembly inventory was picked up by the customer, the sub-lines would simply remake what was consumed. There was some danger, however, that this process assumed the schedule was leveled to some extent.

The danger was overcome by identifying low-volume subassemblies needed for the day's production. These items were built early in the shift to create availability regardless of the timing need.

The other three subassembly areas were located next to the assembly line and were producing to the line's demand. The problem in these three areas was that excess, finished sub-inventory was waiting for consumption. The implementation team decided to reduce the in-process sub-inventory by identifying the maximum required. When the maximum number of units was reached, sub-operators were to stop production and straighten their work areas. This process change was a short-term fix. The long-term countermeasure would be implemented after the rearrangement in July.

Mistake-proofing. The implementation team then shifted its focus to mistake-proofing. Data was unavailable, so the team talked with each operator to discover quality problems. Lean manufacturing requires data-driven decision-making. With little or no data available, operators' experience was used (a marginal substitute). Two primary problems surfaced—a particular painted part was delivered to assembly with a significant amount of scratches, and electronic components malfunctioned during the test process. The first problem could be controlled internally. The second problem required that purchasing employees and line operators meet with suppliers and request sufficient countermeasures to ensure the supplier processes were mistake-proofed. Supplier quality improved almost immediately.

The implementation team looked for a solution to the scratches. The parts were removed from the paint conveyor and packed in a barnhart box. The fill quantity was around 150 pieces, or more than a two-day supply. Operators unloading the parts approached the process with care. Scratches were not introduced during the loading process. Material was then watched as it was delivered to the assembly line. During transportation, the material moved inside the container and the metal-to-metal movement introduced the scratches. Possible solutions were reviewed. Placing packing into returnable dunnage was considered, but it was viewed as too expensive. It was recommended that each part be separated from surrounding parts by a sheet of reusable thin poly material. Trial results showed no defects at assembly, so poly-sheet separation became part of the packing process for these painted parts. Mistake-proofing accomplished the objective.

Visual management. The implementation team turned its attention to visual management—the final area of the standardized work phase. After attending a presentation on the visual factory system, each team leader was given the responsibility of creating a visual environment in their area. The implementation team provided coaching as team leaders reviewed their areas. A consensus was reached that most visual management would be delayed until after the department was reconfigured in July.

Data from the improvements during the standardized work phase are shown in Figures 19-9 and 19-10. As expected, the largest gains were in categories 6.0 (in-station quality) and 7.0 (Just-in-Time). All other categories showed improvements as well.

The Pull Systems Phase

Pull system implementation required a significant amount of planning. The trigger was to be pulled the first day back from the July shutdown, so all systems had to be ready at that time. With line reconfiguration scheduled to occupy most of the shutdown period, the tools of the pull system were to be put into place during the final weekend.

During the pull system phase, the implementation team focused first on developing a "plan for every part." Assistance was requested from the materials group to determine the daily usage of the 2,000-plus parts used on the main assembly and five subassembly lines. A container team was created to determine the types and sizes required for various applications, along with the flow racks to be used. The number and type of containers were defined based on physical size and daily requirements. The priority was to use returnable containers for as many parts as possible. As it turned out, only 12 parts would not fit into the containers. One area of contention was that the container team selected flow racks without casters because the price was about $100 less. In a continuous improvement environment, change is expected and required. Without casters, a forklift was required to move the flow racks. In a lean layout, space is too tight for forklift activity in nondesignated areas. The lack of casters would create problems after the line reconfiguration.

After much discussion, the group elected to have line-side inventories equal to one shift's supply. The TSD consultant unsuccessfully lobbied for a supply equal to half of a shift. Since there was 7,000 ft^2 (650.3 m^2) to be opened with the new layout, there was no problem defining the location of the marketplace. The size of the marketplace was determined by the plan for every part. With fabrication quantities for the assembly line components reset to a five-day supply, marketplace details were identified to handle those requirements.

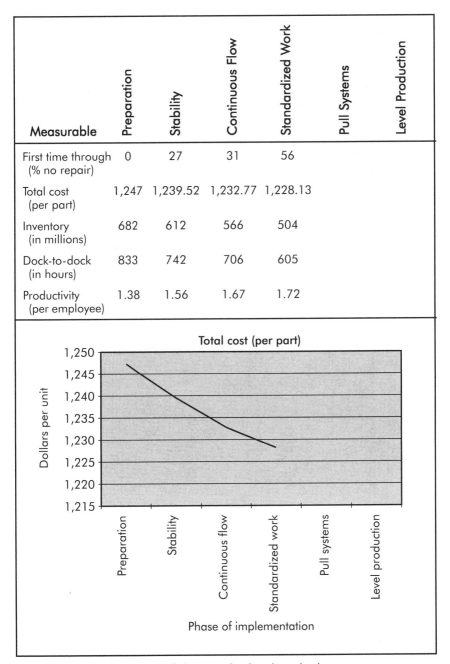

Measurable	Preparation	Stability	Continuous Flow	Standardized Work	Pull Systems	Level Production
First time through (% no repair)	0	27	31	56		
Total cost (per part)	1,247	1,239.52	1,232.77	1,228.13		
Inventory (in millions)	682	612	566	504		
Dock-to-dock (in hours)	833	742	706	605		
Productivity (per employee)	1.38	1.56	1.67	1.72		

Figure 19-9. Lean metrics of the standardized work phase.

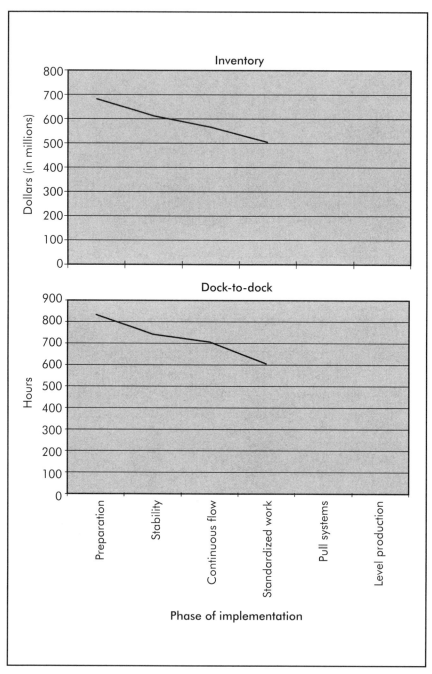

Figure 19-9. (continued).

Brownfield Site Implementation

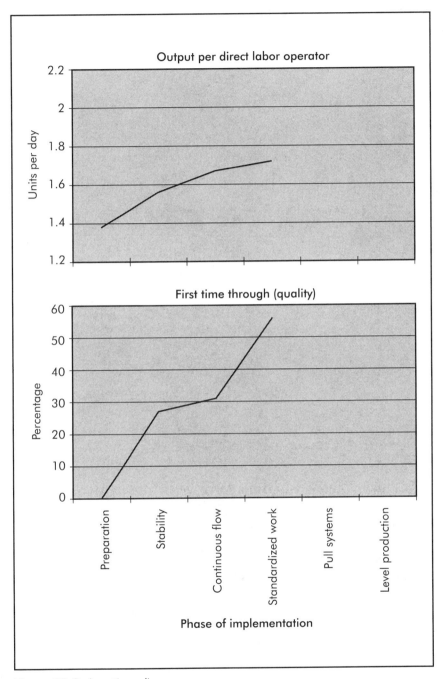

Figure 19-9. (continued).

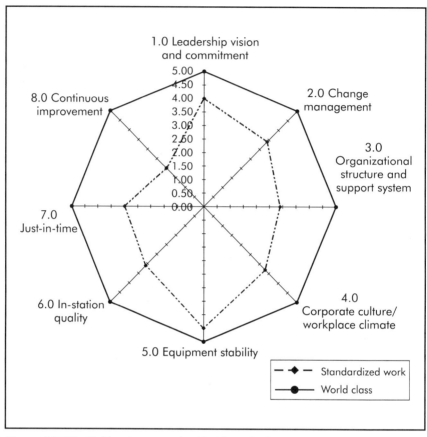

Figure 19-10. Calibrate—standardized work phase.

The maintenance department was asked to create low-cost storage units for the marketplace. The basic design would be 4 × 4 ft (13.1 × 13.1 m), with a slope of 4 in (101.6 mm). It was determined that 2 × 6 in. (50.8 × 152.4 mm) wooden planks and plywood would satisfy the basic requirements. One concern was the roughness of the plywood surface. Imperfections in the wood surface could cause the containers to hang up. Maintenance determined that a Teflon® sheet over the plywood would take care of that issue. Maintenance personnel constructed the bases in June to prepare for the July marketplace launch.

During the next few weeks, the implementation team reviewed the following pull systems checklist and finalized all requirements.

1. Line-side requirements:
   - line-side parts and inventories were defined;
   - returnable containers and flow racks were ordered, and
   - material ID and location cards were created.
2. Marketplace requirements:
   - physical areas were identified;
   - storage bases were in-process;
   - material ID and location cards were created, and
   - Kanban cards were created.
3. Replenishment requirements:
   - were based on usage;
   - delivery cycles were planned;
   - conveyance routes were identified, and
   - material handlers were selected.
4. Pull systems training requirements:
   - were planned for subassembly and assembly operators, as well as material handlers;
   - were completed for production controllers, managers, and the assembly supervisor, and
   - were planned for fabrication supervisors.
5. Fabrication quantity requirement:
   - lot sizes were reduced to one week for assembly parts.
6. Information systems requirements:
   - made sure the order notification system was on track, and
   - terminals were identified for subassembly area usage.

All systems were to be ready for the post-shutdown launch of the new lean assembly line. Figures 19-11 and 19-12 show the improvements made through the pull systems phase.

The Level Production Phase

During the level production phase, the consultant and team leaders focused first on training the appropriate people to use level scheduling (heijunka). Team leaders, production control people, and assembly and operations managers needed to understand the

Measurable	Preparation	Stability	Continuous Flow	Standardized Work	Pull Systems	Level Production
First time through (% no repair)	0	27	31	56	48	
Total cost (per part)	1,247	1,239.52	1,232.77	1,228.13	1,198.96	
Inventory (in millions)	682	612	566	504	328	
Dock-to-dock (in hours)	833	742	706	605	304	
Productivity (per employee)	1.38	1.56	1.67	1.72	1.96	

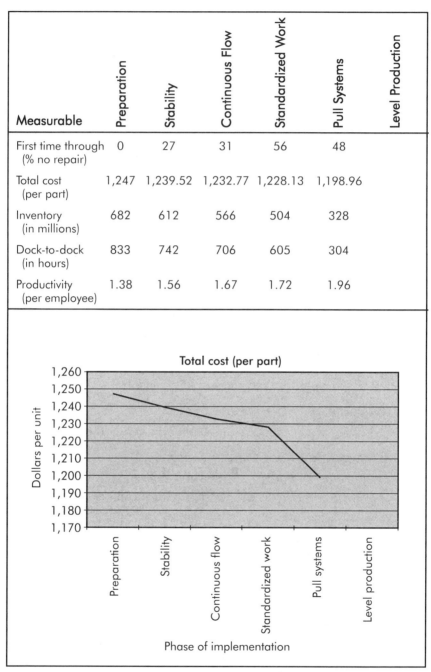

Figure 19-11. Lean metrics of the pull systems phase.

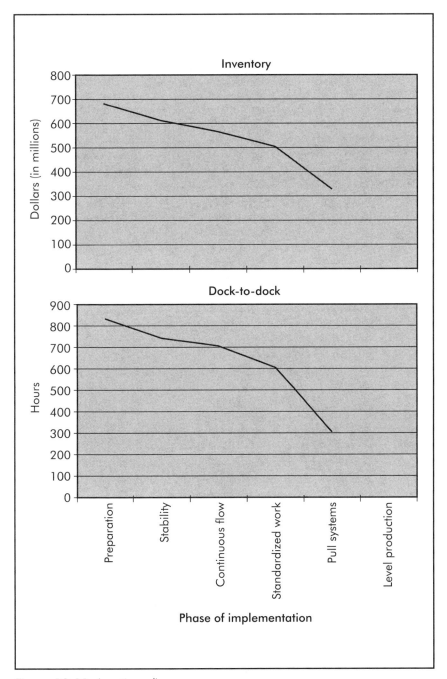

Figure 19-11. (continued).

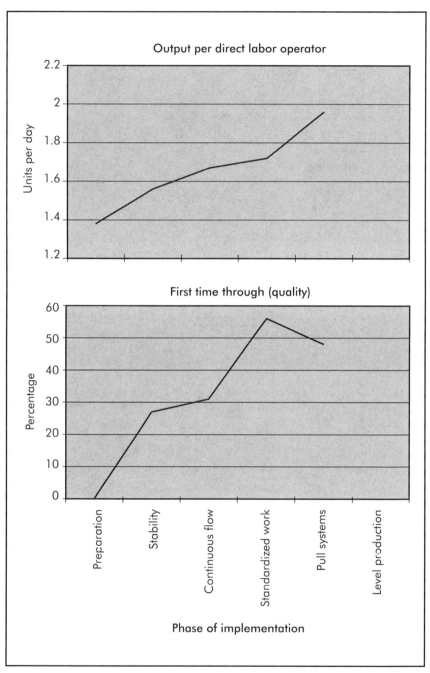

Figure 19-11. (continued).

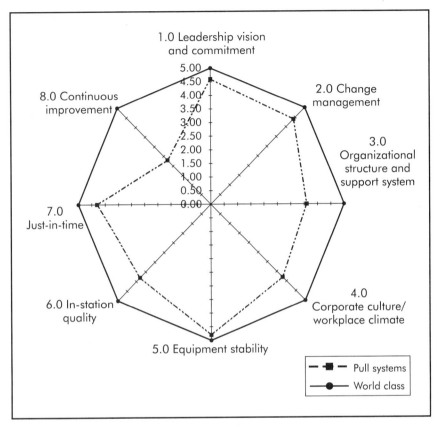

Figure 19-12. Calibrate—pull systems phase.

concepts and requirements of running a leveled schedule. After training, everyone understood the importance of the process and the requirements involved.

Production control personnel studied data on order requirements and developed the cycle process. The assembly line produced six different models, of which one model required extra time in subassembly area three. As it turned out, the special model could only be sequenced to every sixth product. Given that constraint, production controllers began sequencing the assembly schedule before the actual line move occurred.

Industrial engineers spent the weeks leading up to the line change explaining to operators how their jobs would change with the lean layout. Reducing direct labor jobs from 47 to 31 would obviously change the job content. Senior employees in this union shop were given the opportunity to bid on jobs in other areas. Only five chose to do so.

Line reconfiguration during plant shutdown. It was mid-July and the temperature in the facility was hot. With the help of outside contractors, the plant's maintenance group replaced the power-and-free conveyor with a roller conveyor, relocated all air and electrical connections, reinstalled fans above the workstations, and completed all other reconfiguration tasks.

The two local members of the implementation team established the marketplace and the line-side inventory locations, and the inventory manager led moving parts from batch containers to returnables. The marketplace and line-side inventories were filled with available components. Fabrication had not met the target of a five-day supply of all materials in the marketplace.

Launching the Lean Assembly Line

Everything had been planned and executed perfectly and production was going to be extremely smooth the first day of the new lean assembly line. That turned out not to be the case.

As the fourth unit was loaded onto the line, the indexing station developed a cycling problem and shut down. No more units could be started on the line until the station was fixed 1.2 hours later. It broke down four more times that first day. In a flow process, one downed station shuts down the whole assembly line.

A new rule was developed that called for one-piece flow—only one unit between each operation. When flow started again after the indexing stations were fixed, employees at subassembly area number two worked at an extremely slow pace, creating a starved condition for the next operation. The slow pace was primarily a protest against the reduced number of people on the sub-line. In a flow process, one lagging station slows down the entire assembly line.

The testing operation was now manned with five people instead of eight, and some of them were new to their jobs. The test area had been set up for flow, with each person doing a part of the testing versus the previous method of testing the complete machine. Operators were not sure how to identify a problem when it surfaced. Once a team leader helped them identify the problem, a team leader from the supplying area was called to fix it. In most cases, wire harnesses needed to be replaced. This indicated that quality problems still existed with some purchased components. In a flow process, one supplier quality problem shuts down the whole assembly line.

In another case, a fabricated part was formed backwards, preventing its use at line station two. The current material and the rest in the marketplace had the same problem. The fabrication supervisor was required to produce more, and the new parts were received an hour-and-a-half later. In a flow process, internal quality defects shut down the whole assembly line.

Another problem developed in the loading process—a unit was sent down the line without its bar code being scanned. Subsequent subassembly operations did not have the correct parts ready for the unit. An error-proofing suggestion had been made earlier by the consultant to tie the scanning process to an inhibitor bar built into the conveyor. When the bar code was scanned, the bar would drop and allow the unit to pass. The bar was installed on Tuesday, and the electronic interface was completed in a week.

The total production output for the first day was one unit, versus a target of 75 units. Some comments overheard at the end of the day were:

> "Easiest day of work I ever had." —operator
> "Lean failed, let's rip up the line and put it back the way it was!" —supervisor
> "This process is ridiculous." —team leader
> "These problems must be fixed." —vice president of manufacturing
> "Is that what they thought you meant by one-piece flow?" — consultant's wife

All of the problems that surfaced that first day (Monday) had been predicted, but effective countermeasures had not been put in place. Needless to say, the sense of urgency skyrocketed.

Real countermeasures were put in place over the next few days, and many of the problems were solved—some temporarily (with 100% inspection), and others were submitted to a root-cause investigation that concluded with permanent fixes. Production increased on Tuesday to 25 units, 45 units on Wednesday, 52 units on Thursday, and 58 units on Friday. These numbers were well below the target of 75, but getting better every day.

Production eventually leveled off at the pre-lean goal of 65 units per day, still below the 75-unit mark for which the capital appropriation had been approved. Initially, overtime reductions did not occur for various reasons—management practices and fabrication problems were the significant problems, but the primary issue was the lack of integrated support from upper managers. By the end of the year, minimal overtime was being expended to make 65 units per day.

The vice president of manufacturing, who had used TSD on an implementation in another state, drove this particular implementation. The president of the company was not supportive and did not involve the other management disciplines in the implementation. The reduction in quality costs, inventory costs, and the direct labor manpower reduction/reassignment still resulted in a project savings of over $670,000 annually for a six-year period for the assembly line that converted to lean. This was in addition to the one-time $470,000 cash-flow savings from reduced inventory. The lean assembly line accounted for $65M in sales and $47M in costs—savings were 2.3% for the first year, then 1.4% for years two through six. During the implementation process, a new company president with a lean background was hired. He would bring all functions into alignment with lean.

The assessment results following level production are shown in Figures 19-13 and 19-14. In most categories, the future state was reached. The exceptions were in categories 3.0 (organizational structure and support system) and 8.0 (continuous improvement). In time, these categories improved to the level needed.

CONCLUSION

Brownfield conversions present many different challenges to the lean implementer. Emphasis must be placed on careful planning and assessment, a phased implementation plan, and using metrics that support the desired changes. The case study showed some of the ways that a planned strategy could go wrong when executed in a real-world setting. Without careful planning, however, even more

Measurable	Preparation	Stability	Continuous Flow	Standardized Work	Pull Systems	Level Production
First time through (% no repair)	0	27	31	56	48	74
Total cost (per part)	1,247	1,239.52	1,232.77	1,228.13	1,198.96	1,195.32
Inventory (in millions)	682	612	566	504	328	198
Dock-to-dock (in hours)	833	742	706	605	304	270
Productivity (per employee)	1.38	1.56	1.67	1.72	1.96	2.09

Figure 19-13. Lean metrics of the level production phase.

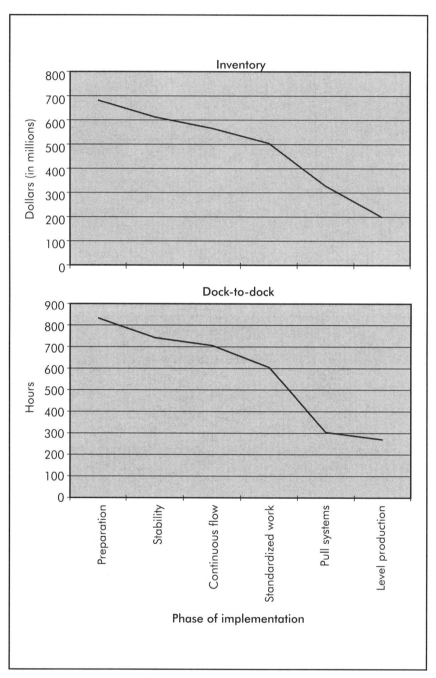

Figure 19-13. (continued).

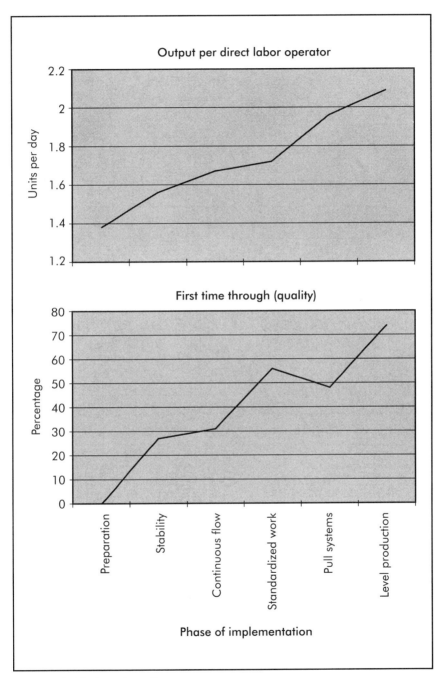

Figure 19-13. (continued).

Lean Manufacturing: A Plant Floor Guide

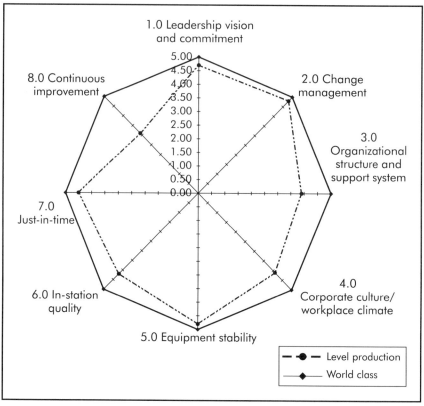

Figure 19-14. Calibrate—level production phase.

chaos is possible and a failed implementation is a real possibility. The consultant's five-phase implementation model provides a template for sequencing the many different activities and changes that are required. Often, different phases of implementation are occurring in different areas of a plant at the same time. Managing the changes becomes a priority for those responsible for the implementation. Finally, the importance of measuring progress and demonstrating the business value of the implementation is critical during the difficult time of transition. TSD's Calibrate and the lean metrics presented in this book (Chapter 3) provide a solid basis for demonstrating that lean does work.

Index

5-minute rule, 32-36 (Figures 2-3 through 2-6)
5-S strategy, 262, 459
5-Why, 19, 39-40, 289, 360-362

A

adjust phase, 373
Andon board, 26

B

batch sizes, 309-310
broadcast system, 397-399 (Figure 17-11)
brownfield site, 256-257, 445
    business case, 455-456
    case study, 447-484
    change management, 446-447
    implementation, 457
    lean launching the line, 482-484
business case, 105, 455-456
    costs, 121-122, 145 (Table 5-5)
    customer satisfaction, 122-123
    employees, 123-125
    equipment maintenance, 118-120
    examples, 141-151
    goals, 106-107
    implementation costs, 132-141 (Table 5-3)
    intangible benefits, 107-108
    inventory, 114-116
    labor, 116-118
    material handling, 120-121
    pilot area, 131-132
    planning, 109-113
    safety, 125-126
    savings, 126-131 (Table 5-2), 135 (Table 5-3), 149 (Table 5-6), 152 (Table 5-7)
    stakeholders, 106, 158

C

Calibrate™, 450, 464 (Figure 19-6), 469 (Figure 19-8), 476 (Figure 19-10), 481 (Figure 19-2), 488 (Figure 19-14)
    current state, 450-452 (Figures 19-1 and 19-2)
    future state, 453-454 (Figure 19-3), 456 (Figure 19-4)
call parts, 425
case study, 447-484
catchball, 225
change, 195-196
change management, 157, 446-447
    developing support, 163
    failure, 158-159
    resistance, 166-170
    roles and responsibilities, 165-167 (Figure 6-1)
    shared vision, 161

stages, 160-165
success factors, 159-160
Charlie's Coil Factory™, 449-450
Check phase, 371
commitment, 196
communication, 30, 32-36 (Figures 2-3 through 2-6)
competition, 195-196
continuous flow, 178-179, 186-187, 189-190, 192, 254, 460, 464-468 (Figure 19-7)
continuous improvement, 295, 400-402
conversion, 447
core values, xxii-xxiv
costs, 121-122, 145 (Table 5-5)
countermeasure, 39, 364-367, 413, 483
culture, 182, 239, 407-408, 446, 450
current state, 310-311, 316-317, 450-452 (Figures 19-1 and 19-2)
current state mapping, 9-10 (Figure 1-4), 73, 76 (Figure 4-1), 452 (Figure 19-2)
   analyzation, 83-87 (Figure 4-7)
   mapping team, 74
   maps, 74-82 (Figures 4-2 through 4-6)
customer, 23-24, 381

D

data collection, 58, 111 (Table 5-1)
   out-of-standard condition, 68-69
   sheets, 65-68
displays and controls, 266-273 (Figures 11-5 through 11-7)
display card, 377
do phase, 370-371
downtime, 459

E

equipment reliability, 326
error proofing, 273-274 (Figure 11-9), 277, 283-284 (Figures 12-1 and 12-2)
   5-Why, 289
   checking methods, 286-287
   defects versus errors, 279
   high volumes, 284
   location, 279-285
   phases of, 285-287 (Figure 12-3)
   red flag conditions, 280
   steps, 287-289
   training, 280
   zero defects, 290-291 (Figure 12-4)
external elements, 312-314 (Figure 14-1)

F

failure mode and effect analysis (FMEA), 38
five-phase implementation model, 173, 175-183, 457-458
   continuous flow, 178-179, 186-187, 189-190, 192, 254, 460, 464-468 (Figure 19-7)
   culture, 182
   definition, 175-177
   lean elements, 184-194
   level production, 180-181, 188, 190, 193, 477, 481-482, 485-487 (Figure 19-13)
   management, 182
   pull production, 180, 187-188, 190, 193, 457, 472, 476-480 (Figure 19-11)
   people system, 193-194
   preparation, 174

quality system, 184
stability, 177-178, 185-186, 189, 191-192, 254, 277, 458-463 (Figure 19-5)
standardized work, 179, 187, 190, 192-193, 295, 458, 465, 469-475 (Figure 19-9)
successful implementations, 183-184
flow, 3-4 (Figure 1-1)
  balanced operations, 8-9
  cellular operations, 7 (Figure 1-2)
  creation of, 14, 97-98
  defined, 3-5
  discrepancies, 18-20
  goals and standards, 6-9
  impact, 5-6
  one-piece flow, 7, 9, 15-16 (Figure 1-7)
  production orientation, 8
  pull systems, 7
  results, 21-22
future state, 453-454 (Figure 19-3), 456 (Figure 19-4)
  mapping, 83, 87-98, 99-103 (Figures 4-17 and 4-18), 456 (Figure 19-4)

G

gemba (go and see), 48-49
Ginder, A. P. and Robinson, L. J., 325
greenfield site, 256-257, 417
  approval, 420
  changeover, 429 (Figure 18-3)
  concepts, 418
  construction, 430-432
  design, 418-430
  lean considerations, 430-432
  lean layout, 421, 423-424 (Figure 18-2)

  needs, 418
  ramp-up, 441-442
  startup, 436-441
  training, 432-435
group communication board, 65-66 (Figure 3-3)

H

heijunka, 100
holding area, 262-263
hoshin kanri, 20, 219

I

implementation, 41-42, 227-230, 247-252 (Figures 10-1 and 10-2), 254-256, 248-249, 310-315, 457
  costs, 132-141 (Table 5-3)
  teams, 166
inspection standards, 24-27 (Tables 2-1 and 2-2)
internal elements, 312-314 (Figure 14-1)
inventory, 4-5, 114-116, 326

J

Jidoka (built-in quality), 26, 28 (Figure 2-1), 426
job security, 200-201, 234
Jones, D. and Womack, J., xxi
just-in-time, 175, 375-377

K

Kanban system, 20-21 (Figure 1-10), 264, 377, 383-399 (Figures 17-4, 17-5, and 17-8), 401-402, 411-412

broadcast system, 397-399 (Figure 17-11)
order point, 391-397 (Figure 17-9)
production, 386-391 (Figure 17-6)
removing, 401-402
training, 408-410
urgency tables, 387-391 (Figure 17-7)
withdrawal, 384, 386

L

leaders, 196-197, 204, 210-211, 222, 240-241
leadership, 165, 169-170, 182-183, 200, 213-214, 408
   command and control, 168
   leaders, 196-197, 204, 210-211, 222, 240-241
   participative, 168
lean human resource system, 198-203
   citizens, 200
   ground rules, 203-207
   job security, 200-201, 234
   leadership, 200
   maintaining the relationship, 207-213
   managing people, 213-215
   manpower planning, 202
   monitoring, 199-200
   problems, 198-199
   training, 201
   vision, 198
lean implementation, 73, 326
lean manufacturing model, 89 (Figure 4-9)
lean measurement, 45-50 (Figure 3-1)
   cycle, 60-69
   defining, 49-58
   gemba (go and see), 48-49
   managing, 47-49
   overall equipment effectiveness, 53-54
   performance, 45-47
   process, 38-60 (Figures 3-2a and b)
learn, use, teach, instruct (LUTI), 328
level production, 180-181, 188, 190, 193, 477, 481-482, 485-487 (Figure 19-13)
line workers, 213

M

make signal, 379-380, 383
managers, 201-202, 204-205
manpower, 317-318
mapping process, 449
mass production, xii, 11 (Figure 1-5), 442 (Figure 18-1), 447
mean time between failure (MTBF), 351
mean time to repair (MTTR), 351
metrics, 419-420, 436
mindset, 166, 169, 212-213
mistake-proofing, 471
move signal, 379

N

Nakajima, Seiichi, 325-326

O

Occupational Safety and Health Administration (OSHA), 260
one-piece flow, 7, 9, 15-16 (Figure 1-7)
organizational structure, 37
overall equipment effectiveness, 53-54

P

pareto analysis, 279-280
periodic checks, 26, 30-31 (Figure 2-2)
plan-do-check-act (PDCA), 37-42, 228-229
plant leaders, 448
policy deployment, 20, 190, 207-209, 219
    advantages, 221-222
    alignment, 220
    considerations, 230
    implementation, 227-230
    principles, 220-221
    process, 222-228 (Figure 9-2)
    steps, 223 (Figure 9-1)
preventive maintenance, 18
principles (lean manufacturing), xxvi-xxxi, 176 (Figure 7-1), 220-221, 303-304
problem identification/solving, 37-42, 357, 372 (Figure 16-2)
    5-Why, 360-362
    adjust phase, 373
    check phase, 371
    countermeasures, 39, 364-367
    do phase, 370-371
    evaluation, 367-370 (Figure 16-1)
    implementation, 41-42
    investigation, 360-367
    mentality of, 357-360
    permanent solutions, 41
    plan-do-check-act (PDCA), 37-42, 228-229
    root cause analysis, 39-40
process parameters, 23
production capacity sheet, 298, 301-302 (Figure 13-3)
production system, 199
pull production, 180, 187-188, 190, 193, 457, 472, 476-480 (Figure 19-11)

pull systems, 7, 21, 375-377
    benefits, 402-403
    company culture, 407-408
    considerations, 404-407
    ensuring success, 410-413
    example, 381-383 (Figure 17-3)
    for continuous improvement 400-402
    just-in-time, 375-377
    management, 377-383 (Figures 17-1 and 17-2)
    operational rules, 400
    tools of, 383-399
    training, 408-409

Q

quality, 459
quality assurance, 24
quality feedback system, 23
quick changeover, 301
    benefits, 308-310
    implementation, 310-315
    issues, 315-320
    streamlining activities, 320-322
    time reduction, 307-308

R

red flag conditions, 280
red tagging, 262
return on investment, 135 (Table 5-3)
Robinson, L. J. and Ginder, A. P., 325
robotics, 427
roles and responsibilities, 165-167 (Figure 6-1)
roll-up, 226
root cause, 39-40, 361, 363

S

savings, 126-131 (Table 5-2), 135 (Table 5-3), 149 (Table 5-6), 152 (Table 5-7)
sensei, 83
seven wastes, xxii, 9, 12 (Figure 1-6), 84-85
stability, 177-178, 185-186, 189, 191-192, 254, 277, 458-463 (Figure 19-5)
standardized work, xxvii, 293, 179, 187, 190, 192-193, 295, 458, 465, 469-475 (Figure 19-9)
   benefits of, 293-294
   documents, 298-301
   elements, 295-298
   pre-conditions, 301-303
   production capacity sheet, 298, 301-302 (Figure 13-3)
   purpose of, 294-295
   roles and responsibilities, 303-305
   work sheets, 295, 298-299 (Figure 13-1), 413
steering committee, 166
streamlining, 320-322

T

Takt time, 14, 90-96 (Figures 4-10 and 4-16), 285, 294, 426
temporary measure, 364
Total Productive Maintenance (TPM), 325, 327 (Table 15-1), 451
   definition of, 325-326
   lean application, 352-354
   maintaining, 354-355
   traditional steps, 326-352 (Tables 15-2 through 15-11)
total system cost, xxx
Total Systems Development, Inc. (TSD), 448, 457, 484

Toyota, 6, 84, 94, 175, 183
   Production System, 108
   productive maintenance, 119
   Supplier Support Center, 175
training, 201, 241-246 (Table 10-1), 280, 408-410, 432-435
triggers, 357-358, 406, 424
trust, 196-197

U

u-cells, 16-18 (Figures 1-8 and 1-9), 424
urgency tables, 387-391 (Figure 17-7)

V

Value Stream Mapping, 73, 448
vision, 174, 224
visual aids, 413
visual factory system, 229, 259, 261 (Figure 11-1), 264 (Figure 11-3), 267 (Figure 11-4)
   benefits, 259-260
   displays and controls, 266-273 (Figures 11-5 through 11-7)
   eliminating waste, 260-261
   error-proofing, 273-274 (Figure 11-9)
   examples, 269
   using, 261-266, 274-275
visual management, 471 471

W

waste, xxi, 235, 294, 301
   eliminating, xxix, 234-235
   indicators, 13
   seven wastes, xxii, 9, 12 (Figure 1-6), 84-85
Womack, J. and Jones, D., xxi, 107

work element, 296-298
workers, 196-197
work groups, 233, 239-240
   anatomy, 240-241
   as an approach, 238-239
   assumptions about people, 235-237
   greenfield versus brownfield, 256-257
   implementation, 247-252 (Figures 10-1 and 10-2), 254-256, 248-249
   issues, 254-256
   leaders, 241
   meetings, 246-247
   pilot areas, 253
   steering committees, 252-253
   training, 241-246 (Table 10-1)
   waste elimination, 234-235
work sheets, 295, 298-299
   (Figure 13-1), 413

Z

zero defects, 23-24, 290-291
   (Figure 12-4)